Fungal Biology

Fungal biology has an integral role to play in the development of the biotechnology and biomedical sectors. It has become the subject of increasing importance as new fungi and their associated biomolecules are identified. The interaction between fungi and their environment is central to many natural processes in the biosphere. The hosts and habitats of these eukaryotic microorganisms are very diverse; fungi are present in every ecosystem on Earth. The fungal kingdom is equally diverse, consisting of seven different known phyla. Yet, detailed knowledge is limited to relatively few species. The relationship between fungi and humans has been characterized by the juxtaposed viewpoints of fungi as infectious agents of much dread and their exploitation as highly versatile systems for a range of economically important biotechnological applications. Understanding the biology of different fungi in diverse ecosystems as well as their interactions with the living and non-living is essential to underpin effective and innovative technological developments. This Series will provide a detailed compendium of methods and information used to investigate different aspects of mycology, including fungal biology and biochemistry, genetics, phylogenetics, genomics, proteomics, molecular enzymology, microscopy, and biotechnological applications in a manner that reflects the many recent developments of relevance to researchers and scientists investigating the Kingdom Fungi. Rapid screening techniques based on screening specific regions in the DNA of fungi have been used in species comparison and identification, and are now being extended across fungal phyla. The majorities of fungi are multicellular eukaryotic systems, and therefore may be excellent model systems by which to answer fundamental biological questions. A greater understanding of the cell biology of these versatile eukaryotes will underpin efforts to engineer certain fungal species to provide novel cell factories for production of proteins for pharmaceutical applications. Renewed interest in all aspects of the biology and biotechnology of fungi may also enable the development of "one pot" microbial cell factories to meet consumer energy needs in the 21st century. To realize this potential and to truly understand the diversity and biology of these eukaryotes, continued development of scientific tools and techniques is essential. As a professional reference, this Series will be very helpful to all people who work with fungi and should be useful both to academic institutions and research teams, as well as to teachers, and graduate and postgraduate students. This volume series tracks the continuous developments in fungal biology with the publication of each volume.

More information about this series at http://www.springer.com/series/11224

Tanya E. S. Dahms • Kirk J. Czymmek
Editors

Advanced Microscopy in Mycology

Springer

Editors
Tanya E. S. Dahms
Department of Chemistry and Biochemistry
University of Regina
Regina
Saskatchewan
Canada

Kirk J. Czymmek
North American Applications and Labs
Carl Zeiss Microscopy
Thornwood
New York
USA

ISSN 2198-7777 ISSN 2198-7785 (electronic)
Fungal Biology
ISBN 978-3-319-22436-7 ISBN 978-3-319-22437-4 (eBook)
DOI 10.1007/978-3-319-22437-4

Library of Congress Control Number: 2015951666

Springer Cham Heidelberg New York Dordrecht London

Printed on acid-free paper

Springer International Publishing AG Switzerland is part of Springer Science+Business Media (www.springer.com)

Preface

Fungi have been under a microscope almost since its inception (Zacharias and Hans Jansen 1595), with Hooke (1665) first describing and illustrating *Phragmidium mucronatum* (parasitic rose rust) and the saprophytic *Mucor*, and Malpighi (1675, 1679) documenting a variety of fungi. Every subsequent development in molecular tagging and microscopic instrumentation has impacted mycology. Since 2008, the Nobel Prize in Chemistry has been awarded twice to researchers who have developed advanced tools for microscopic imaging. The most recent was awarded last year for developing super-resolution fluorescence microscopy, which collectively pushed microscopic resolution beyond the diffraction limit, the holy grail of optical microscopy since first defined by Ernst Abbé in 1873. The second involved the discovery and development of the green fluorescent protein (GFP) which has produced a spectral rainbow of fluorescent proteins for tagging and tracking molecules in living cells. This development, revolutionized biological microscopy, and inspired the discovery of mEos fluorescent proteins which have enhanced certain types of super-resolution microscopy. Such advances now allow us to image cells at resolutions that are an order of magnitude better than diffraction limited optical approaches.

This volume is a compilation of the principles underscoring various advanced microscopy methods and how they have been, or have the potential to be, applied to mycology. Chapter 1 begins with Drs. Rosa Mouriño Perez (Centro de Investigación Cientifica y de Educación Superior de Ensenada) and Robby Roberson (Arizona State University) offering a comprehensive overview of the confocal principle, confocal laser scanning microscopy and its application to fungal biology. In some ways, this chapter is a cornerstone of the volume, as so many of the newer sophisticated fluorescence-based methods depend on this technology. Drs. Norio Takeshita from Karlsruhe Institute of Technology and Oier Etxebeste at the University of Basque Country teamed up in Chapter 2 to introduce us to fluorescence-based techniques that have been used to study fungi, including bimolecular fluorescence complementation (BiFC), the so-called four-letter F-words (i.e. FRET, FRAP, FLIM), and novel dyes (mEos) that have been developed for super-resolution methods. This leads us directly to Chapter 3 in which Drs. James Dodgson, Rafael Carazo Salas, Anatole Chessel from the University of Cambridge and Dr. Susan Cox at King's College London describe the various types of recently developed super resolution microscopy techniques, the subject of this year's Nobel Prize, and how they have been used to uncover minute details within fungi both spatially and temporally. In Chapter 4, Dr. Annette Naumann from the Julius-Kühn Institute describes how Fourier transform infrared (FTIR) microscopy is uniquely poised to image and determine the chemical make-up of fungi alone, and in the context of a common substrate, wood. Chapter 5 turns to Drs. Zhiting Liang, Yong Guan, Shan Chen and Yangchao Tian at the National Synchrotron Radiation Laboratory in the University of Science and Technology in China who describe how full-field hard X-ray tomography has been applied for the first time to reconstruct, at the nanoscale, the three-dimensional (3D) structure of yeast, along with the future potential of this method. Chapter 6 by Dr. Yajing Shen from the City University of Hong Kong outlines a completely novel and clever method for the *in situ* characterization of yeast at the nano scale, using environmental scanning electron microscopy (ESEM) and focused ion beam milling (FIB). Some of the methods from Chapter 6 could also find application to atomic force microscopy (AFM), the subject of Chapter 7 by Drs. Cécile Formosa and Etienne Dague from the Centre National de la Recherche Scientifique. In this chapter, Formosa and Dague tell us how AFM can be used to image and quantify the biophysical properties of live yeast and fungi. Finally in Chapter 8, coeditors Dr. Tanya Dahms from the University of Regina and Dr. Kirk Czymmek Carl Zeiss Microscopy, Inc. describe advances in biosensors, 3D imaging, correlative microscopy, and other recent advances in microscopy methods as applied to fungi.

While most of the methods described in this volume have, at least in principle, mycological applications, there are so many open questions that could be answered using the advanced microscopy described herein. Even for microscopes that are commercialized (confocal, FTIR, AFM), they often remain most effective in the hands of microscopy experts, while other more specialized methods are usually only found in a handful of labs (super resolution and ESEM-FIB) or require a trip to a synchrotron source (X-ray tomography). That being said, those researchers who operate or create specialized instrumentation in their lab are always keen to collaborate (instruments seeking applications) and so I would encourage you to make friends with one or more microscopy specialist. Just like correlative microscopy, the fruit of collaborative research is more like the fruit tree rather than simply a collection of fruit.

In Gratitude

KJC expresses his thanks to Dr. Terry Hammill who let me into his electron microscopy course as an undergraduate and triggered my lifelong passion for microscopy and fungi. Also to Drs. Karen Klomparens, Dr. Stanley Flegler and Dr. Joanne Whallon for the opportunity to learn every advanced microscope and technique I could get my hands on. A special thanks also goes to Drs. Rick Howard and Tim Bourett. Numerous great conversations, numerous experiments, tons of fun over the years, always with a critical eye and searching for the best representation of cell morphology and reality. A special thanks goes to Drs. Seogchan Kang and Dr. Jeff Caplan, for their long-term and ongoing fruitful collaborations. Finally, my gratitude to Drs. Hye-Seon Kim and Dr. Carissa Young who both spent days optimizing and imaging until the result was as good as it could get.

TESD is grateful to Arthur G. Szabo who taught me all I know about lasers and spectroscopy and Linda Johnston who first introduced me to AFM. Dr. Susan Kaminskyj is the person who guided me into mycological microscopy, having met as postdoctoral fellows at Purdue University. Our collaboration, beginning over sips of scotch, has spanned 15 years and has been a constant source of inspiration—this volume would not exist without her! I thank Drs. Vijai Gupta and Maria Tuhoy for the invitation to create this volume as part of the Springer series in mycology, along other collaborative reviews. Their confidence in me and Vijai's constant encouragement has been invaluable for bringing this volume to fruition. I thank Eric Stannard, Springer editor, for his valuable advice, encouragement and flexibility throughout this project and Gina Kahn, assistant editor, who has kept this project on track, offered her copyediting expertise and who wove the chapters together into this volume. Thanks to the University of Regina (UofR) faculty, students and staff, who put up with me during this process, and the UofR for providing the freedom to pursue such a worthy project of this magnitude. TESD's introductory and advanced microscopy undergraduate and graduate students have been a constant source of inspiration—they are my greatest teachers. Finally, I acknowledge that this volume would not have been possible without the support of my partner in life, Douglas Casimel.

Tanya E. S. Dahms and Kirk J. Czymmek

Contents

Contributors

Rafael E. Carazo Salas Genetics Department, University of Cambridge, Cambridge, UK

Shan Chen National Synchrotron Radiation Laboratory, University of Science and Technology of China, Hefei, Anhui, P. R. China

Anatole Chessel Department of Genetics, University of Cambridge, Cambridge, UK

Susan Cox Randall Division of Cell and Molecular Biophysics, King's College London, London, UK

Kirk J. Czymmek North American Applications and Labs, Carl Zeiss Microscopy, LLC, Thornwood, NY, USA

Etienne Dague CNRS, LAAS, Toulouse, France

Université de Toulouse; LAAS, Toulouse, France

Tanya E. S. Dahms Department of Chemistry and Biochemistry, University of Regina, Regina, Saskatchewan, Canada

James Dodgson Genetics Department, University of Cambridge, Cambridge, UK

Oier Etxebeste Biochemistry II Laboratory; Applied Chemistry; Faculty of Chemistry., The University of The Basque Country, San Sebastian, Gipuzkoa, Spain

Cécile Formosa CNRS, LAAS, Toulouse, France

Université de Toulouse; LAAS, Toulouse, France

CNRS, UMR 7565, SRSMC, Vandœuvre-lès-Nancy, France

Faculté de Pharmacie, Université de Lorraine, UMR 7565, Nancy, France

Yong Guan National Synchrotron Radiation Laboratory, University of Science and Technology of China, Hefei, Anhui, P. R. China

Zhiting Liang National Synchrotron Radiation Laboratory, University of Science and Technology of China, Hefei, Anhui, P. R. China

Rosa R. Mouriño-Pérez Department of Microbiology, Center of Scientific Research and Higher Education of Ensenada, Ensenada, B. C., Mexico

Annette Naumann Federal Research Centre for Cultivated Plants, Julius Kühn-Institute, Berlin, Germany

Molecular Wood Biotechnology and Technical Mycology, Forest Botany and Treephysiology, Georg-August-University Göttingen, Büsgen-Institute, Göttingen, Germany

Robert W. Roberson School of Life Sciences, Arizona State University, Tempe, AZ, USA

Yajing Shen Department of Mechanical and Biomedical Engineering, City University of Hong Kong, Hong Kong, P. R. China

Norio Takeshita Department of Microbiology, Karlsruhe Institute of Technology, Karlsruhe, Baden-Württemberg, Germany

Faculty of Life and Environmental Sciences, University of Tsukuba, Tsukuba-shi, Ibaraki, Japan

Yangchao Tian National Synchrotron Radiation Laboratory, University of Science and Technology of China, Hefei, Anhui, P. R. China

Chapter 1
Confocal Laser-scanning Microscopy in Filamentous Fungi

Rosa R. Mouriño-Pérez and Robert W. Roberson

1.1 Introduction

To answer many questions about how a fungal cell functions, it is necessary to examine cellular and subcellular events such as the behaviors and locations of specific organelles and proteins. For this purpose, there are a variety of fluorescence-imaging modalities that are commonly used, such as wide-field microscopy, laser-scanning confocal microscopy (LSCM) (Czymmek et al. 1994; Pawley 2006; Sheppard and Shotton 1997; Diaspro 2001), deconvolution microscopy (Panepinto et al. 2009; Swedlow 2003; Woyke et al. 2002), spinning-disk confocal microscopy (Ichiwara et al. 1999; Thorn 2010), two-photon microscopy (König 2000; Diaspro 2001; Feijo and Moreno 2004), and super-resolution microscopy methods (Huang et al. 2009; see Chap. 3). Confocal microscopy was developed in the 1950s (Minsky 1988). Though a working microscope was produced in 1955, the technological advances required to construct a practical research instrument delayed bringing it to market until the mid-1980s. Confocal microscopy has since become a widely used technique for imaging an extensive range of fluorescently labeled biological specimens, including fungi. Over the years, major improvements have been made in the technology applied in recording and displaying confocal images with high resolution and in determining the availability of more photo-stable fluorophores. Initially, confocal microscopy applications were used mainly in fixing images of fluorescently labeled specimens using, for example, immunofluorescence protocols. The imaging of living organisms has become more applicable to a wider range of specimens with the introduction

R. R. Mouriño-Pérez (✉)
Department of Microbiology, Center of Scientific Research and Higher Education of Ensenada, 3918 Carretera Ensenada-Tijuana, 22860 Ensenada, B. C., Mexico
e-mail: rmourino@cicese.mx

R. W. Roberson
School of Life Sciences, Arizona State University, Tempe, AZ, USA

T. E. S. Dahms and K. J. Czymmek (eds.), *Advanced Microscopy in Mycology*, Fungal Biology, DOI 10.1007/978-3-319-22437-4_1

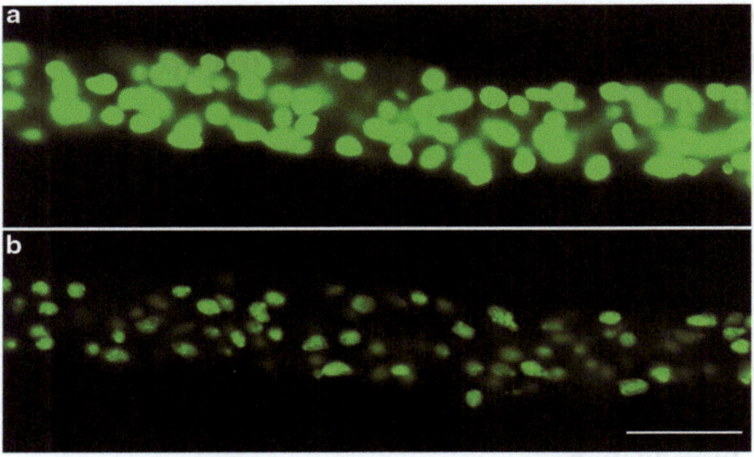

Fig. 1.1 Comparison of (**a**) a wide-field epifluorescence and (**b**) a confocal microscopy image of nuclei labeled with histone H1 fused to GFP in the filamentous fungus *Neurospora crassa*. Scale Bar = 10 µm. (Images by Olga A. Callejas-Negrete)

of fluorescent proteins, use of fluorescent vital dyes, and improved confocal systems (e.g., optics, light path, scanning strategies, and detector electronics) that have reduced photobleaching and phototoxicity. Live-cell imaging techniques provide a means for documenting organelle and macromolecular dynamics with increased spatial and temporal resolutions. Other advantages of confocal microscopy are the experimental approaches for multiple fluorescent labels and multidimensional microscopy.

The "optical section" is the basic image data obtained from a confocal microscope and refers to its ability to produce images along the z-axis (through the sample) of fluorescently-labeled specimens without resorting to physical sectioning of the sample (Fig. 1.1) that have higher contrast and resolution than images produced using a standard wide-field epifluorescence microscope (Conchello and Lichtman 2005; Chandler and Roberson 2009). Optical sections are produced in LSCM by scanning a region of interest of the specimen point-by-point with a focused laser beam, and using a pinhole aperture in front of the detector to remove out-of-focus fluorescent signal from above and below the focal plane of interest. Optical paths of confocal microscopy are designed in such a way that the laser beam is focused within the specimen and the resultant fluorescent signal is refocused at a pinhole in front of the detector (thus, the two focal points, making it confocal). The power of this approach lies in its ability to clearly image structures at discrete Z levels within an intact biological specimen.

1.2 The Confocal Microscopy System

The confocal microscopy system includes integrated optical and electronic devices such as a research grade microscope, multiple sources of coherent light (laser beams), a confocal scan head, a computer with one or two monitors and software for image acquisition, processing and analysis (Fig. 1.2a). The essential part of the LSCM system is the scan head that directs input from one or more laser light sources, sets of filters that select excitation and emission wavelengths, a scanning mechanism, one or more pinhole apertures, and one or more photomultiplier tubes (PMT) as detectors for different fluorescent wavelengths (Fig. 1.2b).

Confocal microscopy was developed to remove out-of-focus light (haze) from the objective focal plane. How does it work?

First, the specimen must be illuminated with a specific wavelength of light to excite the fluorophore. Laser beams are used to illuminate the specimen through the objective lens, which is also a receptacle for the specimen-emitted light; thus, this lens functions as both a condenser and an objective. Dichroic mirrors, prisms, diffraction gratings, or tunable filters are used to obtain specific spectral excitation and/or emission signals. In LSCM, the laser beam fills the objective aperture and forms an intense diffraction-limited spot that is raster-scanned quickly from side to side and from top to bottom pixel by pixel across the entire specimen in a process called point-scanning.

As previously mentioned, the light emitted from the specimen is collected by the objective lens and then follows the microscope's optical path, reaching the pinhole aperture in front of the detector required to produce confocal images. The pinhole aperture is placed at a precise distance from the detector where the image-forming train of light is focused (Fig. 1.3; Chandler and Roberson 2009). The emitted light from a single optical plane within the specimen will pass directly through the center of the aperture. In contrast, light coming from specimen regions below and above the optical focal plane (i.e., out-of-focus) will focus before or after the aperture and are eliminated from the final image (Fig. 1.3; Paddock 2000, 2008; Foldes-Papp et al. 2003; Amos and White 2003; Pawley 2006; Murray et al. 2007). The size of the aperture, as well as the wavelength of exciting light and the NA of the objective lens will determine the resolution of the specimen's optical plane (Z thickness) that can be visualized, which is usually much smaller than the depth of focus for the objective lens (Chandler and Roberson 2009).

There are also alternative ways of collecting emitted light from the specimens when we are interested in a specific wavelength range ("user definable"). Emitted light can be separated into spectral bands using the method called spectral imaging, achieved by utilizing prisms, diffraction gratings, or tunable filters to obtain a specific spectral emission signal (Haraguchi et al. 2002; Czymmek 2005). Each fluorophore has a spectral fingerprint that can be obtained by collecting the emitted light intensity in a user-defined and specific total range that is acquiring images every 10 nm wavelength with a 500–650 nm spectral range. By plotting the intensity values in each image, it is possible to obtain the specific emission curve for the fluorophore of interest. It is possible to get the emission curve for as many fluorophores as needed. This information can be used to separate each spectra in multiple labeled specimens by a process called linear unmixing (Hiraoka et al. 2002; Czymmek 2005). Using spectral imaging it is possible to separate closely overlapping fluorophores and autofluorescence (Lin et al. 2009; Knaus et al. 2013).

Excitation and emission light paths are separated by a dichroic mirror or acousto–optic beam splitter before the light passes through the pinhole aperture to reach

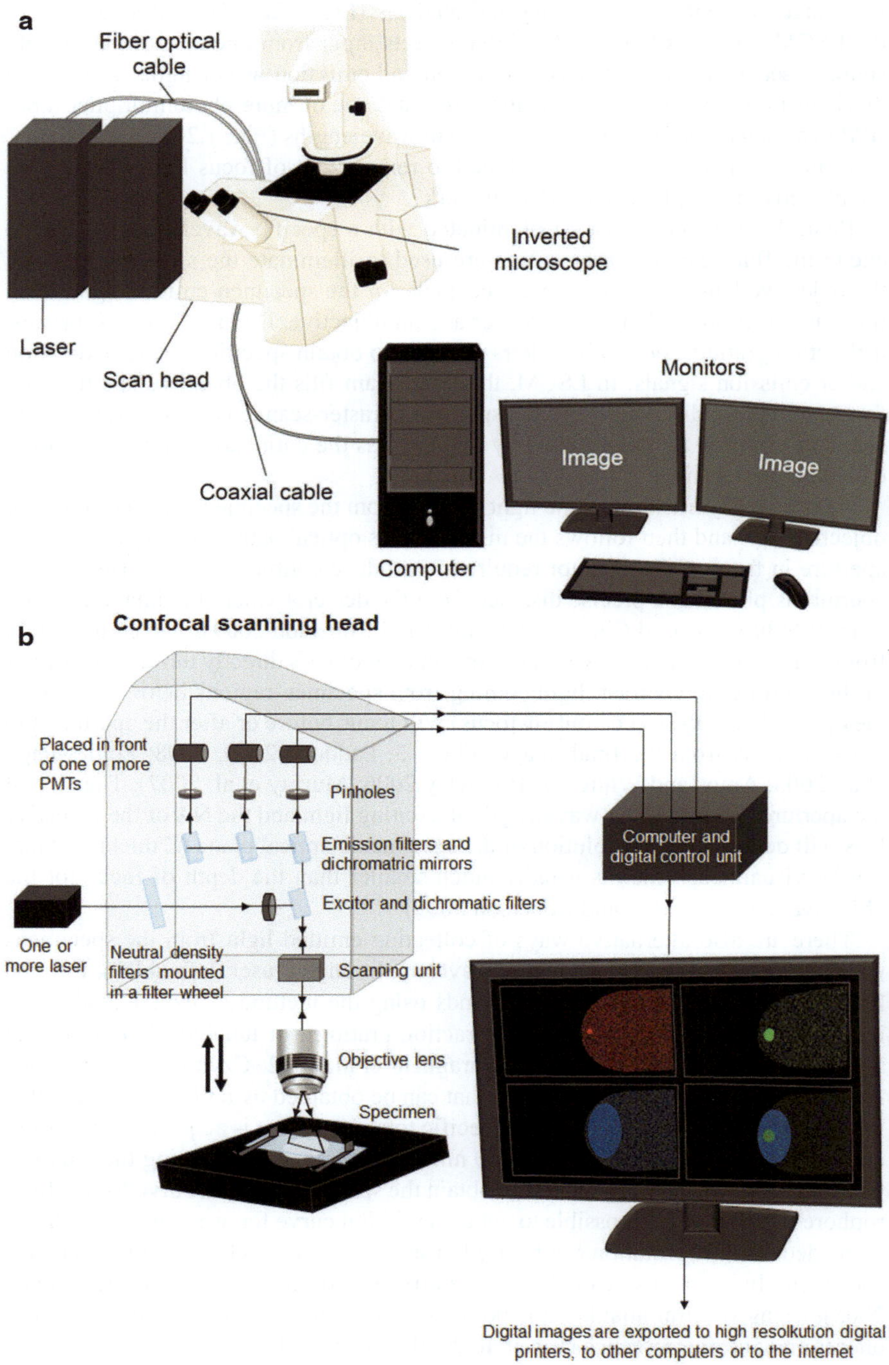

Fig. 1.2 Typical components of a laser-scanning confocal microscopy system. **a** Includes a research-grade microscope, upright or inverted, a scan head, laser beams, computer, and monitors. **b** Diagram of the confocal scanning head. It shows the laser input, the excitation filters and the emission filters, galvano mirrors and detectors. (Artwork by Arianne Ramirez-del Villar)

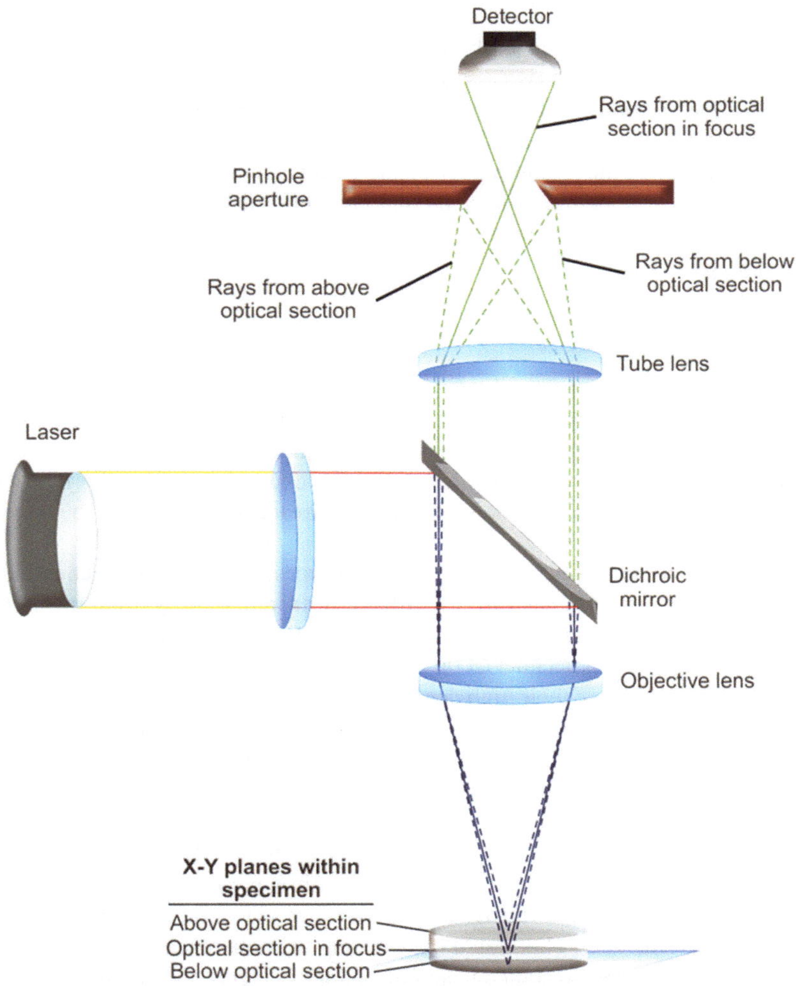

Fig. 1.3 Diagram of the light path in a confocal microscopy system. Pinhole aperture is in front of the detector in order to eliminate out-of-focus light. The image recorded therefore comes from one thin optical plane within the specimen (*solid line*) and the light above and below is blocked (*discontinuous line*). (Artwork by Fausto M. Villavicencio-Aguilar)

the scan head. The filtered light arrives at a galvano-scanning mirror that is constantly oscillating, and sends the signal to the detecting PMT or a more photon-sensitive GaAsP detector at an image plane in a wide-field fluorescence microscope (light path: filters-pinhole-Galvano mirror-detection device). The detectors do not perceive an image per se, but produce a voltage that corresponds to the intensity of incident light (photons) and the software digitizes the signal and displays it as an image on the monitor (Fig. 1.2b; Paddock 2000, 2008; Foldes-Papp et al. 2003; Amos and White 2003; Pawley 2006; Murray et al. 2007).

Each recorded image is an x–y point collection in a specific plane (z) within the specimen. To generate an image of the whole specimen, it is necessary to record an entire "stack" along the z-axis (Z stack) and use them to reconstruct the structure of the specimen with the aid of software that combines planes using mathematical algorithms (Chandler and Roberson 2009). Generally, the software is provided by and is proprietary to the manufacturer of the confocal system, though manufacturers allow export into common formats that can be used for third-party rendering/analysis software such as ImageJ.

1.3 Lasers

Confocal microscopy employs several types of lasers (diode lasers, diode-pumped solid state lasers, and noble gas ion lasers containing argon and helium–neon) to obtain the wavelengths required for fluorophore excitation across the ultraviolet and visible spectra, and to guarantee sufficient fluorescence intensity at the confocal pinhole (Table 1.1).

In the past, highly coherent and intense sources of light were generated by gas lasers (Svelto and Hanna 1989) that have been replaced by solid state lasers having superior performance, price, and longevity (Byer 1988; Wokosin et al. 1996; Kurtsiefer et al. 2000; Koechner 2006). Two important features of lasers used for advanced microscopy are their high degree of temporal coherence or monochromaticity (close to single wavelength) and the spatial coherence (the ability to produce a tight spot). The coherence length is the propagation distance over which a laser maintains a specified degree of coherence (Leith and Upatnieks 1963, 1964; Ackermann and Eichler 2007).

The light intensity profile of a laser beam is Gaussian. Thus, to have an even distribution of light intensity reach the objective, it is necessary to expand the excitation source such that it overfills both the field of view and condenser aperture. Laser speckle needs to be eliminated by reducing the coherence length of the laser beam to eliminate the interference from out-of-focus defects and at the image plane to maximize image resolution (Pawley 2010).

Table 1.1 Lasers used for confocal microscopy

Laser type	Wavelength (nm)				
Ultraviolet-argon	351, 364				
Argon-ion		458, 477, 488	515		
Krypton		488	568	647	
Helium–neon			543	633	
Diode	405	440, 473, 488	539, 559	638, 647	
Diode-pumped solid state	355, 405		532, 561	660, 671	1064
Titanium/sapphire					700–1000

1.4 Multidimensional Data Sets

An outstanding feature of confocal microscopy is the possibility of recording images in several dimensions, either sequentially or simultaneously; space (x, y, and z), time (Fig. 1.4) and emission wavelengths (channels), including non-confocal transmitted light images (Fig. 1.5). Thus, multidimensional images at different depths, time points, and/or multiple wavelengths can be presented as movies, including all or partial information. Different fluorophores can be imaged in a single multi-labeled specimen using a confocal system that combines multiple simultaneous excitation or conventional single-excitation sources. The limitation in the number of fluorophores imaged using the confocal approach is the ability to spectrally discriminate the emission spectra.

Multidimensional images collected by LSCM are usually in register with one another, facilitating their 3D display by computer reconstruction of the Z-series (Fig. 1.6) and the combination of multiwavelength images. Live cells are commonly imaged using time-lapse methods with the related data usually displayed as a time-lapse sequence directly from the confocal microscopy images (see below) (Waterman-Storer 1997).

1.5 Assessing Image Quality and the Performance of the Confocal System

The quality of a confocal image is determined by its spatial resolution, resolution of light intensity (dynamic range), signal-to-noise ratio (S/N), and temporal resolution (Pawley 2010).

Spatial resolution of a system is based on the minimum distance within two points can be distinguished as individual (Chandler and Roberson 2009), and which can be determined quantitatively. The numerical aperture (NA) of an objective and the illumination wavelength (lambda, λ) are crucial for spatial resolution. The minimum resolvable distance d between two points in the x, y plane can be approximated as (Pawley 2010):

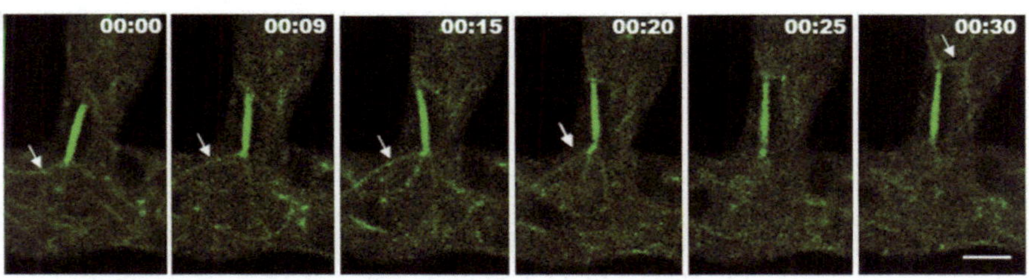

Fig. 1.4 Time lapse of a mitotic spindle in *Neurospora crassa* labeled with β-tubulin fused to GFP. *Arrows* point to astral microtubules. Time in min:sec. Scale bar = 10 μm

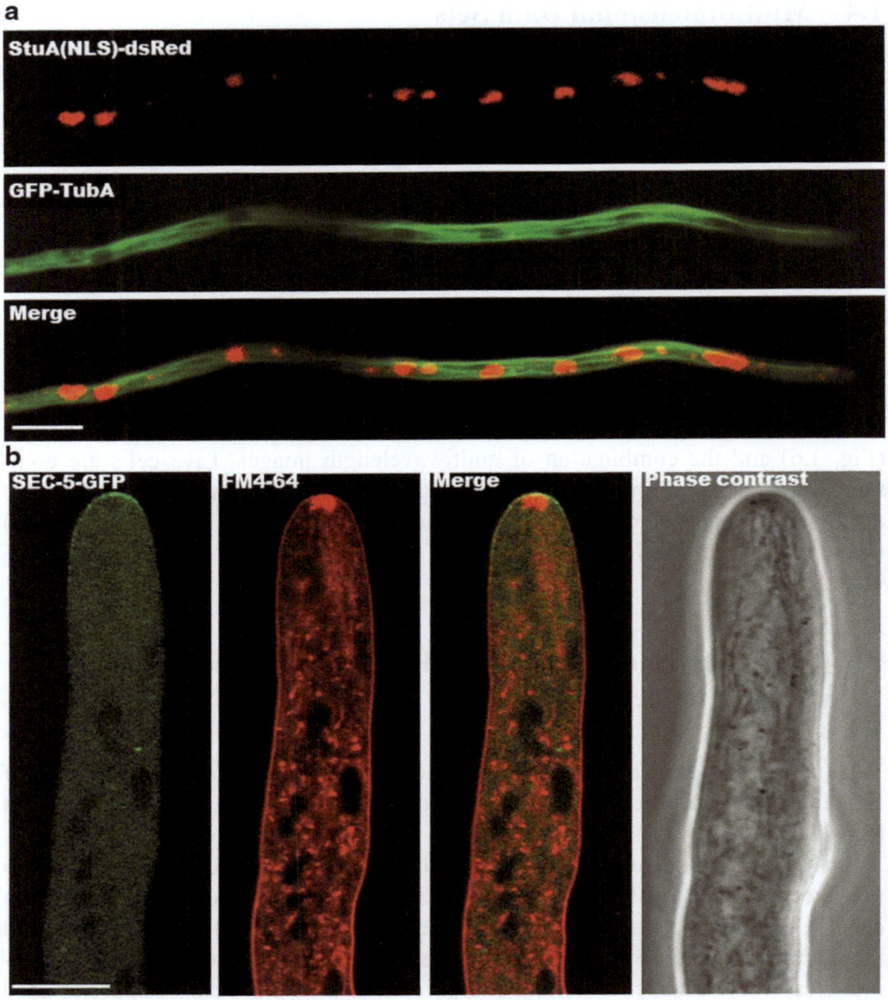

Fig. 1.5 Hyphae multilabeled with fluorescent proteins and vital dyes. **a** Panel 1, single hypha of *Aspergillus nidulans* showing nuclei labeled with StuA(NLS)-dsRed. Panel 2, same cell with microtubules labeled with GFP-TubA. Panel 3, merger of images in panels 1 and 2. Scale bar = 10 μm. (Images by Raphael Manck). **b** Panel 1, single hypha of *Neurospora crassa* labeled with SEC-5-GFP (exocyst component). Panel 2, membranes stained with the lipophilic vital dye FM4-64. Panel 3, merger of images in panels 1 and 2 and panel 3 phase contrast image. Scale bar = 10 μm. (Images by Alejandro Beltran and Meritxell Riquelme)

$$d_{x,y} \approx \frac{0.4\lambda}{\text{NA}}.$$

It is important to use objectives with high NA when the goal is to increase the resolution, since NA is inversely proportional to *d*. Since NA is equal to *n* sinθ, where *n* is the refractive index and θ the angular aperture, the use of oil immersion (higher *n*)

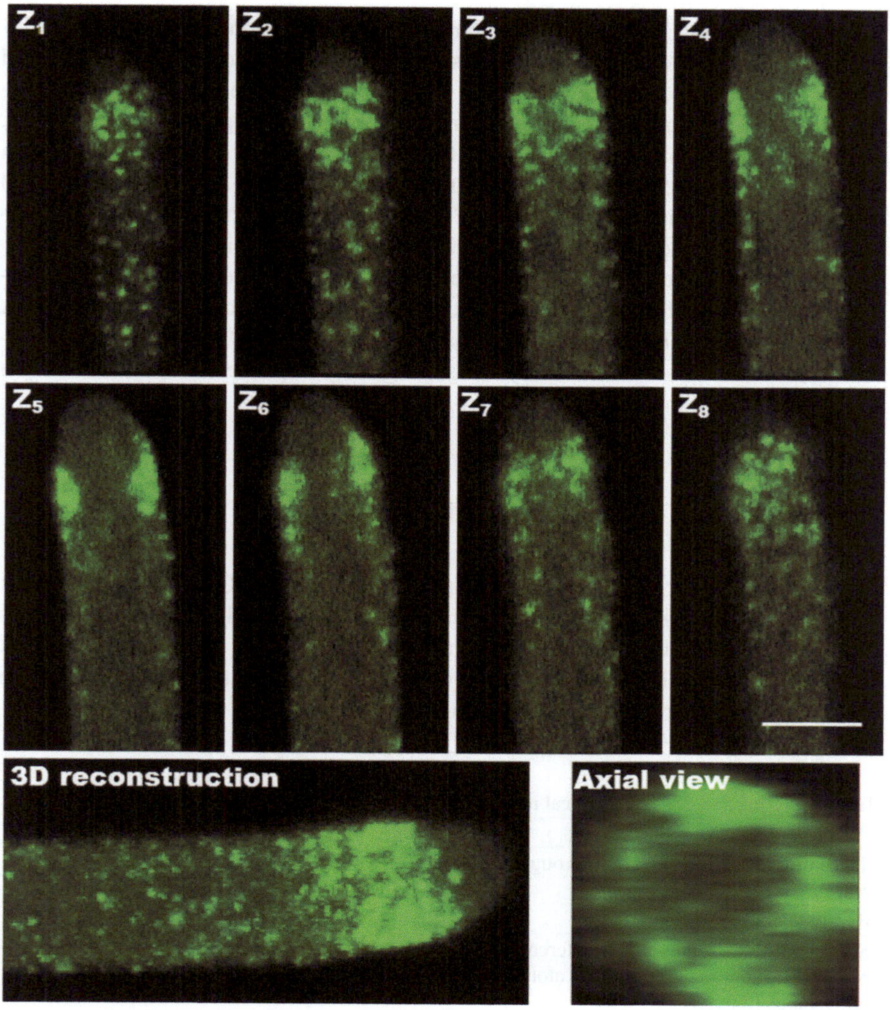

Fig. 1.6 3D reconstruction of the apical and subapical region of *Neurospora crassa* labeled with fimbrin fuse to GFP. (Z_{1-8}) single optical sections of 0.9 μm from top to botton of the hypha. *Bottom panel* in the left shows the reconstruction with all eight optical sections and the *right panel* shows an axial view of the reconstruction. Fimbrin forms a collar around the subapical region. Scale bar = 10 μm

improves resolution. Other factors that can affect spatial resolution are the size of the pinhole (reducing the size of the pinhole improves contrast by eliminating out-of-focus light, but reduces signal collection and thus image intensity), the zoom factor, pixel resolution, and the scan rate (by decreasing the field of view and the scan rate, the amount of light detected increases), all of which can be adjusted using the software (Pawley 2010).

On the detection side, the dynamic range refers to the resolution of light intensity and is defined as the ratio of saturated signal to detector readout noise, calculated in volts or electrons. In other words, the dynamic range can be defined as the number of gray levels that are assigned to an image by the analog-to-digital converter. To attain the maximal potential of the PMT, it is important to acquire images in a dynamic range from no signal (black) to saturating signal (white) (Murphy and Davidson 2013). The signal-to-noise (S/N) ratio, which directly affects image contrast, depends on specimen background and the electronic noise of the system.

Temporal resolution is another aspect for assessing image quality and the performance of a confocal system. Temporal resolution is occasionally a problem, particularly in older generation confocal microscopes. The use of modern resonant scanners and galvo scanner systems has greatly improved temporal resolution issues in confocal microscopy.

1.6 Pros and Cons of Confocal Microscopy

Of the types of microscopy capable of optical sectioning, LSCM has been the most popular. To apply this technique with maximum power and optimal results, it is important to know the associated pros and cons (Table 1.2) to plan experiments that will yield maximal information as answers to our experimental questions (Shotton 1989; Pawley 2010; Hickey and Read 2009).

Table 1.2 Pros and cons of confocal microscopy

Pros	Cons
The ability to optically section through fluorescent objects	Restriction of illuminating wavelengths
Reduction in background signal	Photobleaching
Stacking images representing different focal planes spaced at regular intervals along the optical axis (z-axis)	Relatively slow scan rates
Digitization by an analog to digital converter (transformation of continuous voltage signal into discrete digital steps of light intensity)	Requires associated digital image processor for optimum usefulness
Permits unusually clear examination of thick, light scattering objects	Use of a small confocal aperture reduces overall image intensity
Obtain high-contrast images of surface topologies	Out-of-focus structures invisible
Greatly improved S/N ratio	Bleed-through in fluorophores with similar spectra
Electronic adjustment of magnification	Relatively expensive
Reduce blurring of the image from light-scattering	

1.7 LSCM Advanced Techniques

There are different advanced LSCM techniques such as bimolecular complementation (BiFC), fluorescence recovery after photobleaching (FRAP), fluorescence loss in photobleaching (FLIP), fluorescence lifetime-imaging microscopy (FLIM), fluorescence resonance energy transfer (FRET) that are covered in Chap. 2. Other advanced techniques are fluorescence correlation spectroscopy (FCS) and fluorescence cross-correlation spectroscopy (FCCS), and biophysical techniques that provide quantitative measurements of important parameters, including mechanisms of molecular mobilities, and concentrations in living cells. *In vitro*, FCS examines the inherent correlations exhibited by the fluctuating fluorescent signal from labeled molecules as they transition into and out of a specified excitation volume or area. FCCS, an extension of FCS, measures the temporal fluorescence fluctuations coming from two differently labeled molecules diffusing through a small sample volume. It has become an important method for characterizing diffusion coefficients, binding constants, kinetic rates of binding, and determining molecular interactions (Krichevsky and Bonnet 2002; Baciaet al. 2006; Hwang and Wohland 2007; Tcherniak et al. 2009; Digman and Gratton 2009). Tsujia et al. (2011) used FCS and FCCS to track the two- and three-dimensional motion of quantum dot-conjugated prion proteins in cells of *Saccharomyces cerevisiae.* They developed a method that paved the way for tracking single proteins of interest inside living cells through recording and analyzing the stochastic behaviors of individual quantum dots.

1.8 Comparison of LSCM with Spinning-disc, Deconvolution and Two-Photon Methods

Over the past 70 years, a variety of confocal technologies similar to LSCM have been developed. Spinning-disk confocal (Ichiwara et al. 1999; Nakano 2002), deconvolution (Swedlow 2003), and two-photon confocal microscopy (König 2000; Diaspro 2001; Feijo and Moreno 2004) allow imaging living cells at high spatial resolution, in addition to optical sectioning.

Spinning-disk confocal microscopy is based on the use of the Yokogawa-Nipkow disc that is a combination of two discs, one with an appropriate pattern of pinholes of a fixed size (Nipkow) and the other containing the same pinhole pattern but with a microlens in each pinhole that focuses illuminating light to reduce diffraction-induced spread of the beam (Yokogawa) (Wang et al. 2005). The discs spin synchronously at 6000–8000 rpm, effectively creating a simultaneous illumination of the entire region of interest. Each pattern of pinholes returns to a given spot once per cycle, illuminating each spot for a longer amount of time overall thus reducing the excitation energy needed to illuminate a sample (Nakano 2002; Wang et al. 2005). Since excitation of any given spot is intermittent, phototoxicity and photobleaching of the specimen are reduced and temporal resolution enhanced, which is advantageous for live samples that are delicate (Nakano 2002). Instead of the PMT detectors of LSCM, spinning-disc confocal microscopes use electron-multiplying charge-coupled device (EMCCD) detectors to record images, in addition to enhancing the temporal resolution (Nakano 2002). Excellent examples of the use of spinning-disc confocal microscopy in studying fungal cell biology include the relationship between dynamic microtubules and mitotic spindle poles with mitochondrial positioning in fission yeast (Yaffe et al. 2003), *Ustilago maydis* effector production during plant infection driven by long-distance endosome trafficking (Bielska et al. 2014), the clathrin- and Arp2/3-independent endocytosis in *Candida albicans* (Epp et al. 2013), the regulation of mitosis by the NIMA kinase present in spindle pole bodies of *Aspergillus nidulans*, the chronological appearance of each component of the kinase cascade (Shen and Osmani 2013), and microtubule dynamics during mitosis in *Aspergillus nidulans* (Szewczyk and Oakley 2011); in this study, they used spinning-disk confocal microscopy because it allowed rapid image capture. As microtubule dynamics are extremely rapid, it is necessary to use a fast acquisition system.

One of the major causes of image degradation in wide-field fluorescence microscopy is blurring caused by signals emanating from the out-of-focus fluorescence that would normally be eliminated by the confocal pinhole. This can also be overcome using a computer-based method commonly known as deconvolution microscopy in which deconvolution algorithms either subtract or reassign out-of-focus light. A variety of algorithms are now commercially available for 2D or 3D deconvolution (Wallace et al. 2001; Swedlow 2003; Nasse and Woehl 2010), most easily classified as non-restorative and restorative methods. The former improves contrast by removing out-of-focus light from focal planes, and the latter reassigns light to its proper place of origin. Such methods can be advantageous over 3D confocal microscopy since light is not discarded but rather reused. Images or time series for deconvolution microscopy are recorded by wide-field epifluorescence microcopy, and depending on the detection optics, the spatial and temporal resolution can be very high (Wallace et al. 2001; Swedlow 2003; Nasse and Woehl 2010). Deconvolution microscopy was used successfully in investigating the events of mitosis in *Cryptococcus neoformans* (Woyke et al. 2002), localizing the Rho-GTPase RHO-4 during conidiation versus vegetative septation in the filamentous fungus *Neurospora crassa* (Rasmussen and Glass 2007), understanding the components of the exocyst and polarisome in *Ashbya gossypii* (Köhli et al. 2008), studying the dynamic cellular processes such as growth, cell division, and morphogenesis in budding yeast (Rines et al. 2011) and mapping the yeast kinetochore (Haase et al. 2013); they deconvolved and background subtracted 13 planes, 200 nm apart, to make a 3-D reconstruction of the kinetochore of *Saccharomyces cerevisiae* during mitosis.

An additional imaging approach that can be applied to wide-field fluorescence or confocal microscopy is two- or multiphoton excitation. For two-photon excitation, the most commonly used fluorophores have excitation spectra of 350–500 nm, whereas the laser used for excitation is in the range of ~700–1000 nm (infrared). When a fluorophore absorbs two infrared photons simultaneously, it will absorb enough energy to be raised to the excited state ($E = hc/\lambda$ wavelength is inversely proportional to energy). The fluorophore will emit a single photon with a wavelength that depends on the type of fluorophore used (typically in the visible spectrum). The two-photon method typically requires a pulsed femtosecond laser to quickly deliver photons at high density. The spatial and temporal resolution of two-photon fluorescence microscopy is very similar to LSCM and the two are commonly combined; however, the specialized laser makes it more expensive and slightly more difficult to use. Advantages of two-photon excitation include its superior depth of imaging (up to 1 mm) and minimal phototoxicity (lower energy excitation), making the system well suited for specimens of living organisms (Arcangeli et al. 2000; Czymmek 2005; Knaus et al. 2013). Multiphoton microscopy, with its increased depth of field, has made it possible to document *in vivo* the mechanisms of invasion into the *Arabidopsis root* vascular system by *Fusarium oxysporum* (Czymmek et al. 2007), the multiphoton imaging of medically significant fungi such as *Aspergillus flavus*, *Micosporum gypseum*, *Micosoprum canis*, *Trichophyton rubrum*, and *Trichophyton tonsurans* to analyze their distinct autofluorescence characteristics in microcultures in Sabouraud's agar. They showed that multiphoton microscopy is a powerful noninvasive tool in biomedical imaging. Multiphoton imaging advantages such as strong axial-depth discrimination, enhanced image penetration, and reduced photodamage were revealed to be useful in fungal morphological characterization. The authors showed that hyphae and spores are intensely autofluorescent, providing strong contrast. The fungal cell wall, cytoplasm, and septum were clearly seen. Each fungal species had distinct autofluorescence spectral fingerprints, and the analysis of the multiphoton fluorescence fingerprint could potentially help in differentiating fungal species (Lin et al. 2009).

1.9 Fungal Sample Preparation

With confocal microscopy, it is possible to observe fixed and living fungal specimens. Fixed cells are more robust than living cells, but offer less information about cellular dynamics (Chandler and Roberson 2009).

A common method to localize macromolecules in the context of fixed cells is immunocytochemistry, in which antibodies bind with a high degree of specificity to unique structural features (i.e., epitopes) on the surface of the macromolecules. Whole mount single cells (Hoch and Staples 1985; Roberson 1992; Bourett et al. 1998) or sections of sporangia or complex tissues (Hyde and Hardham 1992) can be prepared for immunocytochemistry with each approach requiring that epitope structural integrity be maintained sufficiently for antibody recognition. Typically, samples are chemically fixed in freshly prepared, buffered (e.g., phosphate-buffered saline) formaldehyde (4 %w/v) for 30 min at room temperature, and then rinsed to remove the excess fixative. These samples can be stored at 4 °C for several days, but it is preferable if they are processed immediately. Samples can also be flash frozen using a variety of cryogens, such as cold (−15−85 °C) organic solvents (e.g., methanol, acetone) (Bourett et al. 1998; Riquelme et al. 2002), or liquid nitrogen cooled propane (Lowry and Roberson 1997; McDaniel and Roberson 1998). After freezing, the samples are incubated in an appropriate solvent at −20 °C overnight, warmed to 4 °C, rehydrated in buffer and processed. Chemically fixed whole mount samples must be permeabilized to allow the diffusion of antibodies into and out of cells during incubation periods. Permeabilization is accomplished using enzyme(s) to partially digest the cell wall (e.g., chitinase) and a gentle detergent (triton X-100) to disrupt the plasma and cytosolic membranes. Disruption of the cell wall may be required for cryo-fixed samples, but detergent treatment is typically not required. Localization of the antigen is carried out indirectly by the use of two antibodies; the first (or primary) antibody localizes the antigen (target macromolecule) whereas the second (or secondary) antibody to the primary antibody is associated with the fluorophore. In order to minimize nonspecific antibody binding to the specimen, incubation in dilute, buffered protein (e.g., bovine serum albumin) is recommended. There are two primary advantages of the indirect approach: (1) a single primary antibody can be localized by multiple secondary antibodies, thus, enhancing signal and sensitivity, and (2) secondary antibodies can be conjugated to a wide choice of inexpensive fluorophores that are commercially available. Antibody dilution and incubation times must be determined empirically for each specimen; 1:100 or 1:1000 are typical dilutions. Following incubation with primary and secondary antibodies, samples are washed and mounted on clean glass slides in a volume of glycerol-based antifade reagent (e.g., n-propyl gallate), and overlaid with a glass cover slip. Samples are stored in the dark at 4 °C until viewing, and sample preparation in the dark can prevent photobleaching of sensitive fluorophores (Chandler and Roberson 2009). Figure 1.7 shows examples of multi-labeled (antibodies and DAPI) fixed fungal hyphae. Some examples of confocal microscopy images of immunolocalization are localizing γ-tubulin as a component of the Spitzenkörper and centrosomes in *Allomyces macrogynus* (McDaniel and Roberson 1998), studying the microtubule cytoskeleton in association with vesicular and mitochondrial motility and positioning in the hyphal apex of *Allomyces macrogynus* (McDaniel and Roberson 2000), and identifying microtubules with detyrosinated α-tubulin and their exclusive motors (Kinesin-3 UncA) in *Aspergillus nidulans*. In this study, the authors showed different populations of microtubules, with just one detyrosinated α-tubulin microtubule in *A. nidulans* that is exclusively associated with Kinesin-3 (Zekert and Fischer 2009).

To maintain the health of cells during live-cell imaging, sample fluorophore concentrations are decreased to reduce cytotoxicity and laser irradiation minimized to prevent phototoxicity and photobleaching. Hickey and Read (2003) reported the "inverted agar block method" (Fig. 1.8) to maintain mycelium in a single focal plane and to reduce spherical aberration (Hickey et al. 2005; Hickey and Read 2009). This method involves cutting a ~2 × 3 cm agar rectangle from the margin of a fungal colony, growing it in an agar plate and leaving plenty of space for hyphae to grow (Fig. 1.8). When cells are labeled with fluorescent proteins, the agar block is placed inverted onto a glass coverslip without any liquid. If extrinsic fluorescent dyes are required, they must be added as ~10–20 μl drops of dye-containing liquid

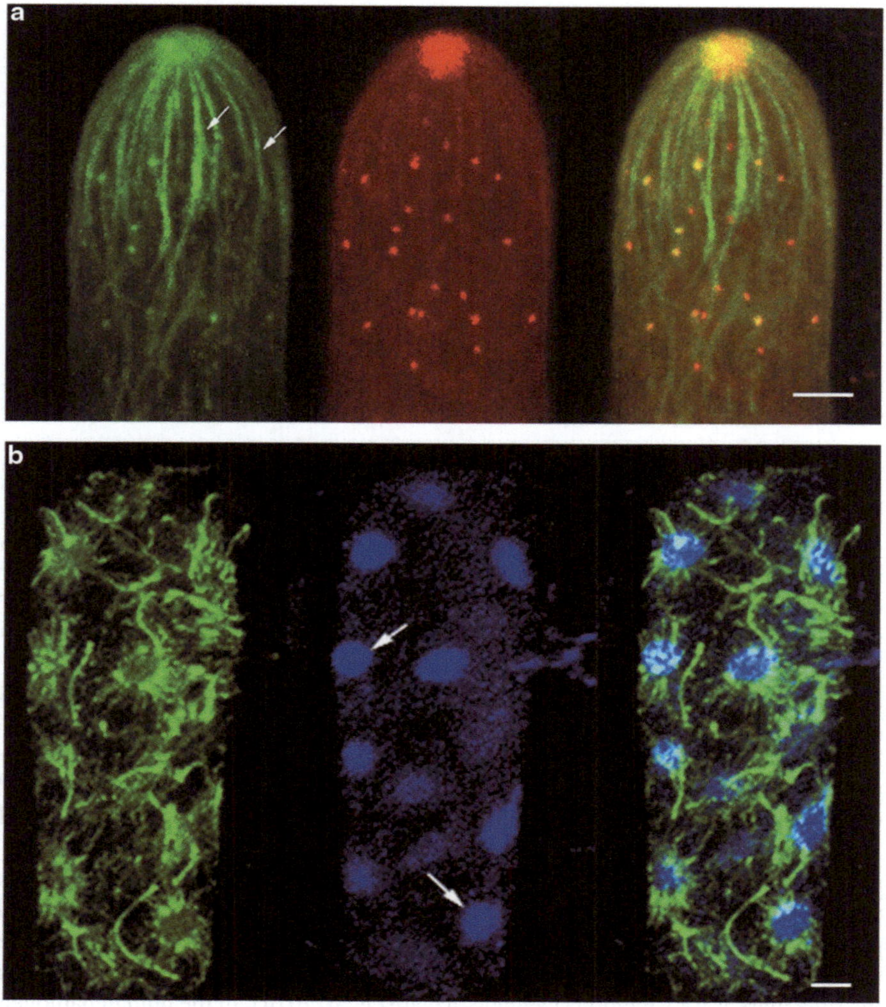

Fig. 1.7 Indirect immunofluorescence labeling of *Allomyces macrogynus* cells. **a** *Panel 1*. A single hyphal tip labeled for visualization of α-tubulin showing abundant microtubules (*arrows*) organized around and emanating from the hyphal apex. *Panel 2*. Same cell as seen in panel 1 showing γ-tubulin localized to a spherical region in the hyphal apex corresponding to the Spitzenkörper, and in the subapical cytoplasm as discrete spots corresponding to centrosomes. *Panel 3*. Merged image of *panels 1* and *2*. Scale bar=2 μm (images by Dennis McDaniels). **b** *Panel 1*. A single zoosporangium undergoing zoosporogenesis labeled for α-tubulin localization showing extensive microtubule arrays. *Panel 2*, same sporangium shown in panel 1 labeled with DAPI for nuclear visualization (*arrows*). *Panel 3*, merged image of *panels 1* and 2. Scale bar=10 μm. (Images by David Lowry)

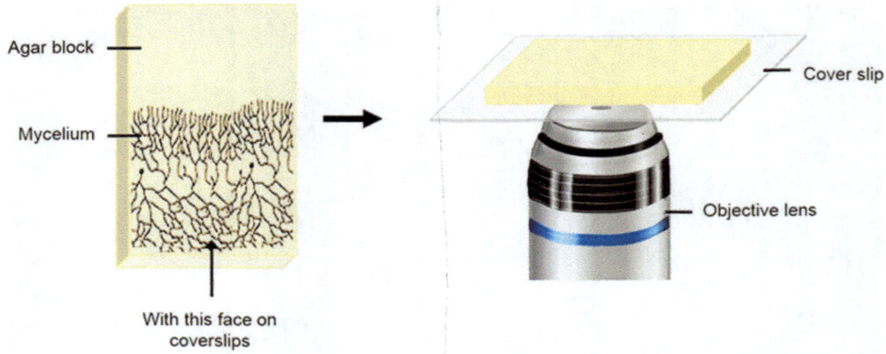

Fig. 1.8 Scheme of the "inverted agar block" developed by Hickey and Read (2003) for living-cell imaging in filamentous fungi

growth medium to the coverslip, before gently placing the agar block (Hickey et al. 2002). Preparation of hyphal cells in this way often results in cessation of growth, thus 10–20 min recovery time is typically required prior to imaging. For multi-labeled live hyphae, fluorescent dyes and proteins can be used in one organism. For instance, we can use GFP and mCherry (or dsRed) (Fig. 1.5) or a GFP labeled strain having FM4-64 stained membranes and calcofluor white stained cell walls. Examples of multi-labeled specimens are shown in the study of the actin cytoskeleton in *Neurospora crassa* during vegetative hyphal growth, where different actin reporters were used to localize the multiple actin-based structures and understand their function that is support and transport. Actin was labeled with the marker Lifeact-GFP (Riedl et al. 2008) and the actin-binding proteins (ABPs: fimbrin, Arp-2, Arp-3, and tropomyosin) that were associated with either GFP or mCherry. This study demonstrated the versatility of actin in both its location throughout the hypha and in its varied associations with ABPs, which regulate the construction-specific actin-based structures (Delgado-Alvarez et al. 2010). Using these methods it was also possible to draw a timeline for actin and ABPs during septum formation, using FM4-64 as the time marker for septation (Delgado-Alvarez et al. 2014). In the human pathogen *Candida albicans*, multi-labeled living hyphae have been useful to study different patterns of localization and distinct dynamic properties of the Spitzenkörper, exocyst, and polarisome components, and the phosphorylation of Exo84 by the cyclin Cdk1-Hgc1 during hyphal extension (Caballero-Lima and Sudbery 2014).

Table 1.3 presents some examples of the different dyes and fluorescent proteins available for fungal cells. Before starting any experiment, it is important to know the excitation and emission spectra of each fluorophore, which may change in the context of the organism. Thus, it is prudent to collect spectral scans of each fluorophore following labeling prior to conducting the confocal experiment. If dual or triple labeling is required, dyes must be chosen that have well-separated emission spectra. Sometimes, it is not possible to resolve the emission signal of fluorophores with similar spectra, and so dyes must be chosen carefully prior to imaging.

Table 1.3 Vital fluorophores and fluorescent proteins used in fugal cells

	Wavelength		
Fluorophore	Excitation	Emission	Target
FM4-64	488	515	Plasma and cytosolic membranes
$DIOC_6$	484	501	Mitochondria, endoplasmic reticulum, and nuclear envelope
DAPI	315	455	DNA and RNA
SYTO dyes	468	533	Nuclei and mitochondrial DNA
Rhodamine 123	507	530	Active mitochondria
DASPMI	475	605	Mitochondria
FM1-43	488	590	Mitochondria
Mitotracker red	644	665	Mitochondria
Rhodamine Phaloidine	540	565	Actin
Calcofluor white	365	490	Cell wall
Solophenyl flavine	440	530	Cell wall
Oregon green	498	526	Vacuoles
Lucifer yellow	428	540	Fluid-phase endocytosis
FITC-dextran	490	520	Fluid-phase endocytosis
Fluorescent proteins			
sGFP	485	510	
eGFP	489	508	
Venus	515	528	
YFP	514	527	
dsRed	556	589	
mCherry	587	610	
EOS	Variable		

Refences for Vital dyes: Johnson et al. 1980; Elorza et al. 1983; Roncero and Duran 1985; Betz et al. 1992, 1996; Knight et al. 1993; Oparka and Read 1994; Vida and Emr 1995; Hoffman and Mendgen 1998; Steinber et al. 1998; Cochilla et al. 1999; Bolte et al. 2004; Fisher-Parton et al. 2000; Read and Hickey 2001; Atkinson et al. 2002; Torralba and Heath 2002; Dijksterhuis 2003; Read and Kalkman 2003; Miyawaki 2003; Bolte et al. 2004; Koll et al. 2001. *References for Fluorescent proteins*: Freitag and Selker 2005; Freitag et al. 2001, 2004; Fuchs et al. 2002; Day and Davidson 2009; Bowman et al. 2009; Leroch et al. 2011.

1.10 Image Processing and Movie-making

Images acquired by LSCM are usually saved in a digital computer format that are easily converted to different file types for manipulation using the proprietary software of the confocal microscope or other commercially or freely available image processing software. It is possible to separate the different channels, time points, and/or the Z-stacks for 3-D reconstruction. Each single image can be converted to any format (e.g., jpg, tif, gif, etc.) depending of the quality desired, but the raw data should always be stored for future processing. Image brightness, contrast, cropping, and/or merging is generally accomplished using image processing software such as Adobe Photoshop®. It is important not to over-manipulate images, thereby, avoiding artifacts representing false information (Chandler and Roberson 2009).

In time-lapse experiments it is possible to make movies that can provide astonishing information about the dynamics of labelled structures or macromolecules in living fungal cells. The number of images is only limited by computer memory, with time intervals generally from 0.5 s to 1 min. A set of images in a time-lapse series can be further processed (contrast changed, brightness changed, cropped, and merged) and saved as a format appropriate for movie-making software (e.g., Adobe Premier®, Quicktime®), most commonly *.avi and *.mov. With the movie-making software, proprietary or not, movies can be accelerated, compressed, edited, and labeled for further presentation. For additional information on these topics see Chandler and Roberson (2009) and Umbaugh (2011).

1.11 Frequently Asked Questions

1.11.1 Do Vital Fluorescent Dyes Work Equally Well for Live-Cell Imaging of All Fungal Species and Cell Types?

Imaging live cells with confocal microscopy that have incorporated vital fluorescent dyes requires using dyes at concentrations that are noncytotoxic and unharmful levels of laser radiation (Hickey et al. 2005); two considerations that can vary between organisms and cell types. Thus, the answer to the question is no; one must consult the literature for previously published protocols or empirically determine what staining and imaging conditions work the best for any given specimen. The bottom line for this and all live-cell imaging is to handle samples gently, making sure they maintain viability and health.

1.11.2 How to Test Cell Viability After Staining?

The use of single or combined dyes can result in cell death, making it important to test the viability of fungal cells following staining prior to live-cell imaging. Staining with fluorescein (live cells) and propidium iodide (dead cells) simultaneously can be used to quantify percentage survival (Oparka and Read 1994). When there is a low percentage of survival, it is important to search for an alternative dye, method, or dye concentration.

1.11.3 How to Assess the Fitness of Living Cells?

The use of certain vital fluorescent probes (e.g., FM4-64), high laser power, or simply contact with the glass cover slip can perturb the cells and interfere with live-cell imaging. It is important to note any cellular perturbations, for example, in hyphae reduced growth rate, swelling or narrowing of the hyphal tip, retraction or disappearance of the Spitzenkörper, and changes in organelle morphology and/or cytoplasmic distribution. Some of these perturbations can be prevented by incubating the sample in an appropriate growth medium for several minutes before performing live-cell imaging (Hickey et al. 2002).

1.11.4 How to Reduce Photobleaching?

Photobleaching is one of the greatest challenges of confocal microscopy, but not as severe as that in wide-field epifluorescence microscopy that uses a different source of illumination that cannot be readily tuned (Shotton 1989). There are differences in both the stability and quenching of the various fluorophores currently available. Undesirable photobleaching can be reduced by minimizing laser intensity or by using longer intervals between scans. The latter solution is not viable if events must be tracked quickly, but it is important to find the balance between scan time, interval between scans, and the time scale of the event to be recorded. Other variables that can influence photobleaching are pinhole size, detector sensitivity, NA of the objective lens, and level of magnification.

1.11.5 What to Do When a Staining Protocol That Normally Works Fails?

A protocol that produces good results can sometimes cease to work, which will relate either to the confocal system or the specimen. It is always good practice to first screen the specimen on a conventional epifluorescence microscope to evaluate sample preparation. If epifluorescence is visible, then a known specimen should also be imaged with the confocal using standard parameters, such as laser power, pinhole diameter, objective lens, field of view, gain, and black level of the detector. The latter is also good practice prior to imaging a new specimen. If there is a problem with the confocal system and the user is unsure of how to proceed, it is best to request technical support from the microscope manufacturer.

1.11.6 How to Deal with Bleed-Through in Multi-labeled Specimens?

For multi-labeled specimens, there can be bleed-through from one detection channel to another that can result from overlapping fluorescence spectra, Förster resonance energy transfer, or a limited number of emission filters. As mentioned before, it is possible to collect the spectral curve of each fluorophore in the context of single-labeled strains and use the information as a reference for a multi-labeled specimen. Some confocal systems have sufficient filters and PMTs to allow the user to define the specific wavelength range of each channel. In this way, the user can select a narrow range associated with the emission maximum for each fluorophore to distinguish the different signals. If such a system is not available, most confocal systems offer software options that deconvolve spectral curves post data collection; alternatively, single detection channels can be collected sequentially.

1.11.7 How to Deal with Background and Autofluorescence?

Autofluorescence occurs naturally in many fungal cells from intrinsic fluorophores (e.g., aromatic amino acids, NADH, FAD, and pyridoxal phosphate) and can be a major source of background interference during imaging. Also certain reagents, such as fixatives (e.g., glutaraldehyde) and culture media components can introduce background fluorescence. Thus, before staining, it is important to observe an unstained specimen at different wavelengths with low laser power and black-level detector settings to identify autofluorescence and background fluorescence. There are specific compounds that can diminish autofluorescence (e.g., sodium borohydride) (Beisker et al. 1987), or fluorophores can be excited at wavelengths longer than (usually visible) autofluorescence (often UV). Another approach to deal with autofluorescence is to remove the signal using digital image-processing techniques such as image subtraction, ideally after having collected images from an unlabelled sample. Spectral imaging methods (see above) can also be applied to minimize autofluorescence. Impure immersion oil used for high-resolution imaging will cause background fluorescence, necessitating fluorescence-free oil for all fluorescence-based microscopies.

1.12 Conclusion

In the past 20 years, there has been a widespread interest in confocal microcopy. This has resulted in considerable improvement in all the optical and electronic components of this bioimaging approach and in the development of a large number of vital fluorescent probes, particularly for imaging live biological specimens. The resolution of confocal microscopy has been proven to be superior to wide-field epifluorescence microscopy. Nevertheless, the full potential of LSCM is yet to come—the idea is to use LSCM for single cell or even single molecule biochemical experiments. A major improvement would be enhancing the ability of LSCM to follow macromolecular intracellular reactions and dynamics, and establish the relationship between specific macromolecules within subcellular compartments.

LSCM is a very powerful tool, versatile, easy to use, and has become a common bioimaging method for many academic and industrial laboratories worldwide. Fungal scientists have found confocal microscopy to be an invaluable tool in addressing broad questions concerning fungal cell biology, from intracellular signaling to cytoplasmic organization and dynamics (Czymmek et al. 1994; Hickey and Read

2009). LSCM requires a wide range of skills, including the ability to operate complex computer software, an understanding of light microscopy optics, and a detailed knowledge of the biology of the system under study. Although, confocal images can be easily obtained, LSCM requires a broad range of skills to take full advantage of the technique. A better understanding of the various technologies involved in LSCM helps to produce more reliable data, and graduate from a "pretty picture" to an "elucidating image" (Hibbs 2004).

References

Ackermann GK, Eichler J (2007) Holography—a practical approach. Wiley, Weinheim

Amos WB, White JG (2003) How the confocal laser scanning microscope entered biological research. Biol Cell 95:335–342

Arcangeli C, Yu W, Cannistraro S, Gratton E (2000) Two-photon autofluorescence microscopy and spectroscopy of Antarctic fungus: new approach for studying effects of UV-B irradiation. Biopolymers 57:218–225

Atkinson HA, Daniels A, Read ND (2002) Live-cell imaging of endocytosis during conidial germination in the rice blast fungus, *Magnaporthe grisea*. Fungal Genet Biol 37:233–244

Bacia K, Kim SA, Schwille P (2006) Fluorescence cross-correlation spectroscopy in living cells. Nat Meth 3:83–89

Beisker W, Dolbeane F, Gray JW (1987) An improved immunocytochemical procedure for high-sensitivity detection of incorporated bromodeoxyuridine. Cytometry 8:235–239

Betz WJ, Mao F, Bewick GS (1992) Activity-dependent fluorescent staining and destaining of living vertebrate motor nerve terminals. J Neurosci 12:363–375

Betz WJ, Mao F, Smith CB (1996) Imaging exocytosis and endocytosis. Curr Opin Neurobiol 6:365–371

Bielska E, Higuchi Y, Schuster M, Steinberg N, Kilaru S, Talbot NJ, Steinberg G (2014) Long-distance endosome trafficking drives fungal effector production during plant infection. Nat Commun. doi:10.1038/ncomms6097

Bolte S, Talbot C, Boutte Y, Catrice O, Read ND, Satiat-Jeunemaitre B (2004) FM-dyes as experimental probes for dissecting vesicle trafficking in lving plant cells. J Microsc 214:159–173

Bourett TM, Czymmek KJ, Howard RJ (1998) An improved method for affinity probe localization in whole cells of filamentous fungi. Fungal Genet Biol 24:3–13

Bowman BJ, Draskovic M, Freitag M, Bowman EJ (2009) Structure and distribution of organelles and cellular location of calcium transporters in *Neurospora crassa*. Eukaryot Cell 8:1845–1855. doi:10.1128/EC.00174-09

Byer R (1988) Diode laser—pumped solid-state lasers. Science 239(4841):742–747. doi:10.1126/science.239.4841.742

Caballero-Lima D, Sudbery PE (2014) In *Candida albicans*, phosphorylation of Exo84 by Cdk1-Hgc1 is necessary for efficient hyphal extension. Mol Biol Cell 25:1097–1110

Chandler DE, Roberson RW (2009) Bioimaging: current concepts in light and electron microscopy, 1st edn. Jones and Bartlett Publishers LLC, Sudbury

Cochilla AJ, Angelson JK, Betz WJ (1999) Monitoring secretory membrane with FM1-43 fluorescence. Annu Rev Neurosci 22:1–10

Conchello JA, Lichtman JW (2005) Optical sectioning microscopy. Nat Methods 2:920–931

Czymmek KJ (2005) Exploring fungal activity with confocal and multiphoton microscopy. In Dighton J, White JF, Oudemans P (eds) The fungal community: its organization and role in the ecosystem, 3rd edn. CRC Press, Boca Raton

Czymmek KJ, Whallon JH, Klomparens A (1994) Confocal microscopy in mycological research. ExpMycol 18:275–293

Czymmek KJ, Fogg M, Powell DH, Sweigard J, Park SY, Kang S (2007) In vivo time-lapse documentation using confocal and multi-photon microscopy reveals the mechanisms of invasion into the Arabidopsis root vascular system by *Fusarium oxysporum*. Fungal Genet Biol 44:1011–1023

Day RD, Davidson MW (2009) The fluorescent protein palette: tools for cellular imaging. Chem Soc Rev 38:2887–2921. doi:10.1039/B901966A

Delgado-Álvarez DL, Callejas-Negrete OA, Gómez N, Freitag M, Roberson RW, Smith LG, Mouriño-Pérez RR (2010) Visualization of F-actin localization and dynamics with live cell markers in *Neurospora crassa*. Fungal Genet Biol 47:573–586

Delgado-Álvarez DL, Bartnicki-García S, Seiler S, Mouriño-Pérez RR (2014) Septum development in *Neurospora crassa*: the Septal Actomyosin Tangle. PLoS One 9(5):e96744

Diaspro A (2001) Confocal and two-photon microscopy: foundations, applications, and advances. Wiley-Liss, New York

Digman MA, Gratton E (2009) Fluorescence correlation spectroscopy and fluorescence cross-correlation spectroscopy. Wiley Interdiscip Rev Syst Biol Med 1:273–282

Dijksterhuis J (2003) Confocal microscopy of Spitzenkörper dynamics during growth and differentiation of rust fungi. Protoplasma 222:53–59

Elorza MV, Rico H, Sentandrew R (1983) Calcofluor white alters the assembly of chitin fibirls in *Saccharomyces cerevisiae* and *Candida albicans* cells. J Gen Microbio 129:1577–1582

Epp E, Nazarova E, Regan H, Douglas LM, Konopka JB, Vogel J, Whiteway M (2013) Clathrin- and Arp2/3-independent endocytosis in the fungal pathogen *Candida albicans*. mBio 4:e00476–e00413. doi:10.1128/mBio.00476–13

Feijo JA, Moreno N (2004) Imaging plant cells by two-photon excitation. Protoplasma 223:1–32

Fisher-Parton S, Parton RM, Hickey PC, Dijksterhuis J, Atkinson HA, Read ND (2000) Confocal microscopy of FM4-64 as a tool for analyzing endocytosis and vesicle trafficking in living fungal hyphae. J Microsc 198:246–259

Foldes-Papp Z, Demel U, Tilz GP (2003) Laser scanning confocal fluorescence microscopy: an overview. Int Immunopharmacol 3:1715–1729

Freitag M, Selker EU (2005) Expression and visualization of red fluorescent protein (RFP) in *Neurospora crassa*. Fungal Genet Newslett 52:14–17

Freitag M, Ciuffetti LM, Selker EU (2001) Expression and visualization of green fluorescent protein (GFP) in *Neurospora crassa*. Fungal Genet Newslett 48:15–19

Freitag M, Hickey PC, Raju NB, Selker EU, Read ND (2004) GFP as a tool to analyze the organization, dynamics and function of nuclei and microtubules in *Neurospora crassa*. Fungal Genet Biol 41(10):897–910

Fuchs F, Prokisch H, Neupert W, Westermann B (2002) Interactions of mitochondria with microtubules in the filamentous fungus *Neurospora crassa*. J Cell Sci 115:1931–1937

Haase J, Mishra PF, Stephens A, Haggerty R, Quammen C, Taylor RM, Yeh E, Basrai MA, Bloom K (2013) A 3D map of the yeast kinetochore reveals the presence of core and accessory centromere-specific histone. Curr Biol 23:1939–1944

Haraguchi T, Shimi T, Koujin T, Hashiguchi N, Hiraoka Y (2002) Spectral imaging fluorescence microscopy. Genes Cells 7:881–887

Hibbs AR (2004) Confocal microscopy for biologists. Springer, New York

Hickey PC, Read ND (2003) Biology of living fungi. British Mycological Society, Stevenage. (CD–ROM)

Hickey PC, Read ND (2009) Imaging living cells of *Aspergillus* in vitro. Med Mycol 47(Supplement 1):S110–S119

Hickey PC, Jacobson DJ, Read ND, Glass NL (2002) Live-cell imaging of vegetative hyphal fusion in *Neurospora crassa*. Fungal Genet Biol 37:109–119

Hickey PC, Swift SR, Roca MG, Read ND (2005) Live-cell imaging of filamentous fungi using vital fluorescent dyes and confocal microscopy. Method Microbio 34:63–87. doi:10.1016/S0580-9517(04)34004-1

Hiraoka Y, Shimi T, Hashiguchi N (2002) Multispectral imaging fluorescence microscopy for living cells. Cell Struct Func 27:367–374

Hoch HC, Staples RC (1985) The microtubule cytoskeleton in hyphae of *Uromyces phaseoli* germlings: its relationship to the region of nucleation and to the F-actin cytoskeleton. Protoplasma 124:112–122

Hoffman J, Mendgen K (1998) Endocytosis and membrane turnover in the germ tube of *Uromyces fabae*. Fungal Genet Biol 24:77–85

Huang B, Bates M, Zhuang X (2009) Super resolution fluorescence microscopy. Annu Rev Biochem 78:993–1016. doi:10.1146/annurev.biochem.77.061906.092014

Hwang LC, Wohland T (2007) Recent advances in fluorescence cross-correlation spectroscopy. Cell Biochem Biophys 49:1–13

Hyde GJ, Hardham AR (1992) Confocal microscopy of microtubule arrays in cryosectioned sporangia of *Phytophthora cinnamomi*. ExpMycol 16:201–218

Ichiwara A, Tanaami T, Ishida H, Shimizu M (1999) Confocal fluorescent microcopy using a Nipkow scanner. In: Mason WT (ed) Fluorescent and luminescent probes for biological activity, 2nd edn. Academic, San Diego, pp 344–349

Johnson LV, Walsh ML, Chen LB (1980) Localization of mitochondria in living cells with rhodamine 123. Proc Natl Acad Sci U S A 77:990–994

Knaus H, Blab GA, Agronskaia AV, van den Heuvel DJ, Gerritsen HC, Wösten HAB (2013) Monitoring the metabolic state of fungal hyphae and the presence of melanin by nonlinear spectral imaging. Appl Environ Microbiol 79:6345–6350

Knight H, Trewavas AJ, Read ND (1993) Confocal microscopy of living fungal hyphae microinjected with Ca^{2+}-sensitive fluorescent dyes. Mycol Res 97:1505–1515

Koechner W (2006) Solid-state laser engineering, vol. 1. 6th, rev. and updated ed. Springer, Berlin

Köhli M, Galati V, Boudier V, Roberson RW, Philippsen P (2008) Growth-speed correlated localization of exocyst and polarisome components in growth zones of *Ashbya gossypii* hyphal tips. J Cell Sci 121:3803–3814

Koll F, Sidoti C, Rincheval V, Lecellier G (2001) Mitochondrial membrane potential and ageing in *Podospora anserina*. Mech Ageing Dev 122:205–217

König K (2000) Multiphoton microscopy in life sciences. J Microsc 200:83–104

Krichevsky O, Bonnet G (2002) Fluorescence correlation spectroscopy: the technique and its applications. Rep Prog Phys 65:251–297

Kurtsiefer C, Mayer S, Zarda P, Weinfurter H (2000) Stable solid-state source of single photons. Phys Rev Lett 85:290

Leith EN, Upatnieks J (1963) Wavefront reconstruction with continuous-tone objects. J Opt Soc Am 53(12):1377–1381

Leith EN, Upatnieks J (1964) Wavefront reconstruction with diffused. J Opt Soc Am 54(11):1295–1301

Leroch M, Mernke D, Koppenhoefer D, Schneider P, Mosbach A, Doehlemann G, Hahn M (2011) Living colors in the gray mold pathogen *Botrytis cinerea*: codon-optimized genes encoding green fluorescent protein and mcherry, which exhibit bright fluorescence. Appl Environ Microbiol 77(9):887–2897. doi:10.1128/AEM.02644–10

Lin S-J, Tan H-Y, Kuo C-J, Wu R-J, Wang S-H, Chen W-L, Jee S-H, Dong C-Y (2009) Multiphoton autofluorescence spectral analysis for fungus imaging and identification. Appl Phys Lett 95:43703–43703. doi:10.1063/1.3189084

Lowry D, Roberson RW (1997) The microtubule cytoskeleton during zoospore formation in *Allomyces macrogynus*. Protoplasma 196:45–54

McDaniel DP, Roberson RW (1998) γ-Tubulin is a component of the Spitzenkörper and centrosomes in hyphal tip cells of *Allomyces macrogynus*. Protoplasma 203:118–123

McDaniel DP, Roberson RW (2000) Microtubules are required for motility and positioning of vesicles and mitochondria in hyphal tip cells of *Allomyces macrogynus*. Fungal Genet Biol 31(3):233–244

Minsky M (1988) Memoir on inventing the confocal scanning microscope. Scanning 10:128–138

Miyawaki A (2003) Fluorescence correlation spectroscopy for the detection and study of single molecules in biology. Bioessays 24:758–764

Murphy DB, Davidson MW (2013) Fundamentals of light microscopy and electronic imaging. 2nd ed. Wiley, Hoboken

Murray JM, Appleton PL, Swedlow JR, Waters JC (2007) Evaluating performance in three dimensional fluorescence microscopy. J Microsc 228:390–405

Nakano A (2002) Spinning-disk confocal microscopy-a cutting-edge tool for imaging of membrane traffic. Cell Struct Func 27:349–355

Nasse MJ, Woehl JC (2010) Realistic modeling of the illumination point spread function in confocal scanning optical microscopy. J Opt Soc Am A 27(2):295–302. doi:10.1364/JOSAA.27.000295

Oparka KJ, Read ND (1994) The use of fluorescent probes for studies on living plant cells. In Harris N, Oparka KJ (eds) Plant cell biology. A practical approach. IRL Press, Oxford

Paddock SW (2000) Principles and practices of laser scanning confocal microscopy. Mol Biotech 16:127–149

Paddock SW (2008) Over the rainbow: 25 years of confocal imaging. Bio Techniques 44:643–648

Panepinto J, Komperda K, Frases S, Park YD, Djordjevic JT, Casadevall A, Williamson PR (2009) Sec6-dependent sorting of fungal extracellular exosomes and laccase of *Cryptococcus neoformans*. Mole Micro 71:1165–1176

Pawley JB (2006) Handbook of biological confocal microscopy, 3th ed. Plenum Press, New York

Pawley JB (2010) Handbook of biological confocal microscopy (Google eBook). Springer, New York

Rasmussen CG, Glass NL (2007) Localization of RHO-4 indicates differential regulation of conidial versus vegetative septation in the filamentous fungus *Neurospora crassa*. Eukaryot Cell 6:1097–1107

Read ND, Hickey PJ (2001) The vesicle trafficking network and tip growth in fungal hyphae. In: Geitmann A, Cresti M, Heath IB (eds) Cell biology of plant and fugal tip growth. IOS Press, Amsterdam

Read ND, Kalkman ER (2003) Does endocytosis occur in fungal hyphae? Fungal Genet Biol 39:199–203

Riedl J, Crevenna AH, Kessenbrock K, Yu JH, Neukirchen D, Bista M, Bradke F, Jenne D, Holak TA, Werb Z, Sixt M, Wedlich-Söldner R (2008) Lifeact: a versatile marker to visualize F-actin. Nat Methods 5:605–607

Rines DR, Thomann D, Dorn JF, Goodwin P, Sorger PK (2011) Live cell imaging of yeast. Cold Spring Harb Protoc 9:pdb.top065482

Riquelme M, McDaniel DP, Roberson RW, Bartnicki-García S (2002) The effect of ropy-1 mutation on cytoplasmic organization in mature hyphae of *Neurospora crassa*. Fungal Genet Biol 37:171–179

Roberson RW (1992) The actin cytoskeleton in hyphal cells of *Sclerotium rolfsii*. Mycologia 84:41–51

Roncero C, Duran A (1985) Effect of calcoflúor white and congo red on fungal wall morphogenesis: in vivo activation of chitin polymerization. J Bacteriol 170:1950–1954

Shen KF, Osmani SA (2013) Regulation of mitosis by the NIMA kinase involves TINA and its newly discovered partner An-WDR8, at spindle pole bodies. Mol Biol Cell 24:3842–3856

Sheppard CJR, Shotton DM (1997). Confocal laser scanning microscopy. IOS Scientific Publishers, Oxford

Shotton DM (1989) Confocal scanning optical microscopy and its applications for biological specimens. J Cell Sci 94:175–206

Steinber G, Schliwa M, Lehmler C, Bölker M, Kahmann R, McIntosh JR (1998) Kinesin from the plant pathogenic fungus *Ustilago maydis* is involved in vacuole formation and cytoplasmic migration. J Cell Sci 111:2235–2246

Svelto O, Hanna DC (1989) Principles of lasers, 3rd edn. Plenum, New York, p. 494

Swedlow JR (2003) Quantitative fluorescence microscopy and image deconvolution. Meth Cell Biol 72:346–367

Szewczyk E, Oakley BR (2011) Microtubule dynamics in mitosis in *Aspergillus nidulans*. Fungal Genet Biol 48(10):998–999

Tcherniak A, Reznik C, Link S, Landes CF (2009) Fluorescence correlation spectroscopy: criteria for analysis in complex systems. AnalytChem 81:746–754

Thorn K (2010) Spinning-disk confocal microscopy of yeast. Meth Enzmol 470:581–602

Torralba S, Heath IB (2002) Analysis of three separate probes suggests the absence of endocytosis in *Neurospora crassa* hyphae. Fungal Genet Biol 37:221–232

Tsujia T, Kawai-Nomaa S, Packb C-G, Terajimaa H, Yajimac J, Nishizakac T, Kinjod M, Taguchia H (2011) Single-particle tracking of quantum dot-conjugated prion proteins inside yeast cells. Biochem Bioph Res Co 405(4):638–643

Umbaugh SE (2011) Digital image processing and analysis. CRC Press, Boca Raton

Vida TA, Emr SD (1995) A new vital stain for visualizing vacuolar membrane dynamics and endocytosis in yeast. J Cell Biol 128:779–792

Wallace W, Schaefer LH, Swedlow JR (2001) A workingperson's guide to deconvolution in light microscopy. Biotechniques 31(5):1076–1078

Wang E, Babbey CM, Dunn KW (2005) Performance comparison between the high-speed Yokogawa spinning disc confocal system and single-point scanning confocal systems. J Microsc 218:148–59

Waterman-Storer CM, Shaw SL, Salmon ED (1997) Production and presentation of digital movies. Trends Cell Biol 7:503–506

Wokosin DL, Centonze VE, Crittenden S, White J (1996) Three-photon excitation fluorescence imaging of biological specimens using an all-solid-state laser. Bioimaging 4:208–214

Woyke T, Winkelmann G, Roberson RW, Pettit GR, Pettit RK (2002) Three-dimensional visualization of microtubules during the *Cryptococcus neoformans* cell cycle and effects of auristatin PHE. Antimicrob Agents Chemother 46:3802–3808

Yaffe MP, Stuurman N, Vale RD (2003) Mitochondrial positioning in fission yeast is driven by association with dynamic microtubules and mitotic spindle poles. Proc Natl Acad Sci U S A. 100:11424–8

Zekert N, Fischer R (2009) The *Aspergillus nidulans* kinesin-3 uncA motor moves vesicles along a subpopulation of microtubules. Mol Biol Cell 20:673–684

Chapter 2
Fluorescence-Based Methods for the Study of Protein Localization, Interaction, and Dynamics in Filamentous Fungi

Oier Etxebeste and Norio Takeshita

2.1 Introduction

Many filamentous fungal species are directly relevant to humans and human activities (Jiang et al. 2013). Many play important roles in industry, agriculture, and medicine (Gibbs et al. 2000). For example, antibiotics or immunosuppressants such as penicillins, griseofulvin, cephalosporin, and cyclosporins are obtained from *Penicillium, Chepalosporium, Tolepocladium,* and Cylindrocarpon species, respectively (Gutierrez-Correa et al. 2012; Kürnsteiner et al. 2002). Mycotoxins such as ochratoxins, aflatoxins, and zearalenone are produced by *Aspergillus, Penicillium,* and *Fusarium* species, while enzymes such as glucose oxidase, pectinase, xylanase, or glycoamylases are commercially produced from *Aspergilli* (Gibbs et al. 2000). Other filamentous fungal species are pathogens of plants and animals, including humans (Jiang et al. 2013). Commonly carried dimorphic (yeast/mold) fungi such as *Candida albicans* or *Candida dubliniensis* and the opportunistic fungus *Aspergillus fumigatus* are the best-known examples of human and animal pathogens that afflict individuals whose immune systems are compromised or dysfunctional (Brand 2012). An unprecedented number of fungal diseases have recently caused some of

N. Takeshita (✉)
Department of Microbiology, Karlsruhe Institute of Technology, Hertzstrasse 16, Geb. 06.40, 76187 Karlsruhe, Baden-Württemberg, Germany
e-mail: norio.takeshita@kit.edu; takeshita.norio.gf@u.tsukuba.ac.jp

Faculty of Life and Environmental Sciences, University of Tsukuba, 1-1-1 Tennoudai Tsukuba, Tsukuba-shi, Ibaraki 305-8572, Japan

O. Etxebeste
Biochemistry II Laboratory; Applied Chemistry; Faculty of Chemistry. The University of The Basque Country, Manuel de Lardizabal 3, 20018 San Sebastian, Gipuzkoa, Spain
e-mail: oier.echeveste@ehu.eus

T. E. S. Dahms and K. J. Czymmek (eds.), *Advanced Microscopy in Mycology,*
Fungal Biology, DOI 10.1007/978-3-319-22437-4_2

the most severe extinctions in wild species ever witnessed and constantly threaten food security (Fisher et al. 2012). The ProMED website (program for monitoring emerging diseases; www.promedmail.org) documented, over the period from 1995 to 2010, an increase from 0.4 to 5.4% in the relative proportion of plant fungal pathogen alerts (Fisher et al. 2012). *Fusarium graminearum, Botrytis cinerea,* and *Ustilago maydis* are three examples of the many filamentous fungal plant pathogens infecting wheat, barley, maize or other plant species (Raffaele and Kamoun 2012; Soanes et al. 2008).

One of the main characteristics enabling filamentous fungi to colonize multiple substrates is their unique growth and developmental behavior as well as their ability to rapidly and efficiently adapt to changing environments. For these reasons, several filamentous fungal species serve as model organisms for understanding basic biological processes. The filamentous ascomycete *Aspergillus nidulans* has been employed worldwide for more than 60 years as a model organism because it is closely related to clinically and economically important *Aspergilli* and it is easily manipulated in the laboratory. A vast array of signaling cascades transduces environmental signals in *A. nidulans* and these pathways control the factors that regulate different morphogenetic programs (see, e.g., Etxebeste et al. 2010). The most characteristic cell type of filamentous fungi is the vegetative hypha. This non-specialized, syncytial (multinucleated) cell is characterized by a continuous polarized (apical) growth mode, mediated at the tip through the addition of new material that is transported from distal regions (Fischer et al. 2008; Harris 2006; Momany 2002; Riquelme 2013; Takeshita et al. 2014). The exposition of *A. nidulans* hyphae to air and light induces the production of conidiophores, asexual reproductive structures bearing thousands of propagules called conidia, which are characterized by low water content and arrested metabolism (Adams et al. 1998; Kües and Fischer 2006). These features permit dispersal and survival under adverse environmental conditions (Gregory 1966). The asexual cycle in *A. nidulans* is normally followed by the production of sexual reproductive structures or cleistothecia, which contain meiospores called ascospores (Pontecorvo et al. 1953; Todd et al. 2007).

The implementation of fluorescence microscopy 20 years ago for the study of filamentous fungi has significantly improved our knowledge of the morphogenetic processes described above. Initial studies followed the subcellular distribution of specific organelles and structures using fluorescent dyes (see, e.g., Fischer and Timberlake 1995; Harris et al. 1994; Momany and Hamer 1997). In 2004, different laboratories standardized the labeling of proteins from *A. nidulans* with fluorescent tags. This was achieved through fusion-polymerase chain reaction (PCR), double-joint PCR (Yang et al. 2004; Yu et al. 2004), or using the GATEWAY cloning systems (Toews et al. 2004), and opened an avenue for *in vivo* systematic analyses of the localization and function of *A. nidulans* proteins. A new generation of fluorescence microscopy techniques have been developed in recent years. A variety of advanced methods (bimolecular fluorescence complementation (BiFC), Förster resonance energy transfer (FRET), fluorescence recovery after photobleaching (FRAP), fluorescence lifetime imaging microscopy (FLIM), photoactivated localization microscopy (PALM), stochastic optical reconstruction microscopy (STORM), etc.) are now available for deeper and more accurate analyses of protein localization, interaction, dynamics, and degradation (Day and Davidson 2009). Some of these techniques have been successfully implemented in filamentous fungi and will be described here. Focusing mainly on *A. nidulans,* this chapter will provide the reader with an overview of advanced microscopy tools. On the one hand, tools for the analysis of protein interaction will be presented, such as BiFC, FRET, and FLIM. On the other hand, microscopy tools for the study of protein dynamics, time-lapse or stream acquisitions, FRAP, and PALM (Chap. 3) will be discussed.

2.2 Localizing Proteins and Their Interactions

The functionality of a protein is dependent on its correct folding and transport to the corresponding subcellular compartment and, additionally, on its interaction with specific partners or auxiliary proteins. Furthermore, some factors or protein complexes transduce signals from one cell compartment to the nucleus to activate appropriate response mechanisms. This section will focus on the advanced microscopy methods currently available for *in vivo* monitoring of protein interactions within filamentous fungal cells.

2.2.1 Bimolecular Fluorescence Complementation (BiFC)

The analysis of protein interactions in living cells (what, when, and where) is significant because protein interactions offer fundamental mechanisms for specific biological sensing, signaling, and regulation. BiFC analysis is a technology typically used to obtain spatial information about protein interactions, enabling their direct visualization in living cells. It is based on the association of two nonfluorescent fragments of a fluorescent protein that are attached to different components of the same molecular complex. Proteins that are postulated to interact are fused to unfolded complementary fragments of a fluorescent reporter protein and expressed in live cells. Interaction between the two target proteins will bring the fragments of the fluorescent protein within proximity, allowing the formation of the native three-dimensional structure of the reporter protein and emission of its characteristic fluorescent signal (Fig. 2.1a; Kerppola 2006). The BiFC assay has allowed the visualization of numerous protein interactions in many different cell types and organisms (Kodama and Hu 2012).

One example of BiFC analysis in *A. nidulans* is the study of interactions between "cell-end markers" (Higashitsuji et al. 2009; Takeshita et al. 2008). The cell-end markers TeaA and TeaR localize at the tip of hyphae and control growth direction (Fischer et al. 2008). TeaA is transported to the hyphal tip by growing microtubules (MT) and it is anchored at the apical membrane by the interaction with the membrane-associated receptor TeaR (Fig. 2.1b). The N-terminal half of YFP (YFP^N) and the C-terminal half of YFP (YFP^C) were fused to the cell-end markers. YFP fluorescence was not detected for strains expressing either YFP^N–TeaR or

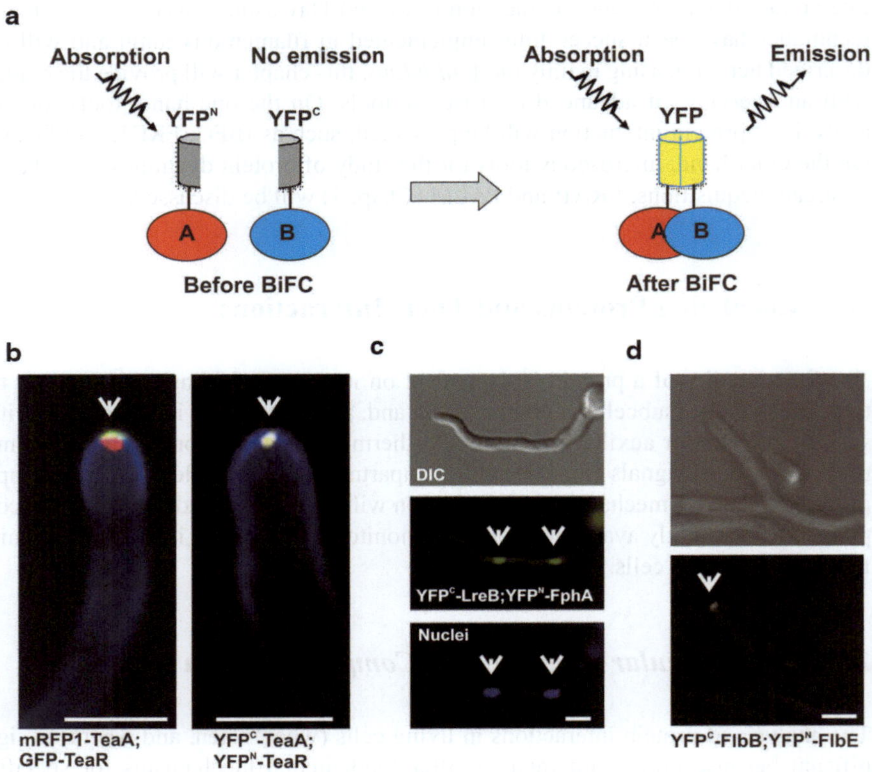

Fig. 2.1 Bimolecular fluorescence complementation (*BiFC*). **a** Schematic representation describing the basis of the technique. Protein A is labeled with the N-terminal half of YFP while Protein B is labeled with the C-terminal half. Formation of a Protein A/Protein B complex brings the YFP fragments within proximity, allowing YFP to acquire its native three-dimensional structure and emit a yellow fluorescence signal. **b, c,** and **d** Examples of protein interactions highlighted with the BiFC technique. The interaction between the cell-end markers TeaA and TeaR is shown in panel **b**, that between light receptors LreB and FphA in panel **c,** and the complex formed by the developmental regulators FlbB and FlbE in panel **d**. Scale bar in all panels = 5 μm

YFP^C–TeaA, but was for the strain expressing both YFP^N–TeaR and YFP^C–TeaA as a single point along the apex (Fig. 2.1b), indicating the site of TeaA and TeaR interaction. Another example of BiFC analysis in *A. nidulans* is the study of interactions between blue (LreB) and red (FphA)-light sensors (Purschwitz et al. 2008), key players in the control of fungal development (Bayram et al. 2010; Rodriguez-Romero et al. 2010). While the interaction was first shown by co-immunoprecipitation, BiFC analysis revealed that their interaction took place in nuclei (Fig. 2.1c) .

The BiFC technique has also been shown to be a valuable tool for the study of the interaction between apically located regulators of fungal development. The transcription factor FlbB interacts with the small filamentous fungal-specific protein FlbE (Garzia et al. 2010) at the tip of vegetative hyphae (Fig. 2.1d; Garzia et al.

2009). The apical localization of both proteins, which depends on this interaction, is essential to induce conidiophore development (asexual reproduction).

BiFC presents two main drawbacks. On the one hand, false positives can occur when noninteracting proteins are present at high concentrations. On the other hand, BiFC is generally not suitable for following dynamic interactions, since the reconstitution of the fluorescent protein is not completely reversible. However, BiFC has been used to study MT polymerase AlpA, which moves along with the MT plus-end towards the hyphal tip. The cell-end marker TeaA is involved in MT guidance to the polarity site through interaction with AlpA. In a strain expressing both YFP^{N}–TeaA and YFP^{C}–AlpA, YFP signals were clearly detected at all hyphal tips (Takeshita et al. 2013). Because AlpA localizes at MT plus-ends, and TeaA accumulates at the hyphal tip cortex, the YFP signal at hyphal tips possibly represents a connection between MT plus-ends and the hyphal tip membrane, suggesting that there is at least a temporary interaction. Although AlpA does not localize at hyphal tips permanently due to MT depolymerization, the YFP signal was nonetheless stable. Tracing interaction dynamics, however, requires other methods such as FRET.

2.2.2 *Förster Resonance Energy Transfer (FRET)*

In living cells, protein interactions play a key role in regulating many signal transduction pathways. FRET is a method that can be used to track dynamic interactions based on energy transfer between two chromophores. A donor chromophore (cyan fluorescent protein (CFP) in Fig. 2.2a), initially in its excited electronic state, may transfer energy to an acceptor chromophore (yellow fluorescent protein (YFP) in Fig. 2.2a) through nonradiative dipole–dipole coupling. The efficiency of this energy transfer is inversely proportional to the sixth power of the distance between the donor and the acceptor, making FRET extremely sensitive to small distances (Tsien 1998). Measurements of FRET efficiency can be used to determine if two fluorophores are within a certain distance of each other, finding wide application in biology and chemistry (Sekar and Periasamy 2003).

One example of a FRET system in filamentous fungi is the dynamic visualization of calcium concentration using the "cameleon" biosensor (Kim et al. 2012). Cameleon consists of calmodulin (CaM) and the CaM-binding module M13 sandwiched between cyan (FRET donor) and yellow (acceptor) fluorescent proteins (Fig. 2.2b; Nagai et al. 2004). In the absence of Ca^{2+}, the excitation of CFP leads to the emission of cyan fluorescence. However, when the CaM module is Ca^{2+} loaded, it interacts with M13 and there is a conformational change that brings CFP within FRET distance of YFP, resulting in yellow fluorescence. This reversible interaction is detected as an increase of the YFP to CFP emission ratio.

Monitoring Ca^{2+} by the cameleon system has been performed in diverse living cells (Allen et al. 1999; Rogers et al. 2007). The use of cameleon in *Magnaporthe oryzae, Fusarium oxysporum,* and *F. graminearum* showed a cytoplasmic Ca^{2+} change in a pulsatile manner, with no discernible gradient between pulses (Kim et al. 2012), which was age and development dependent. Although the number

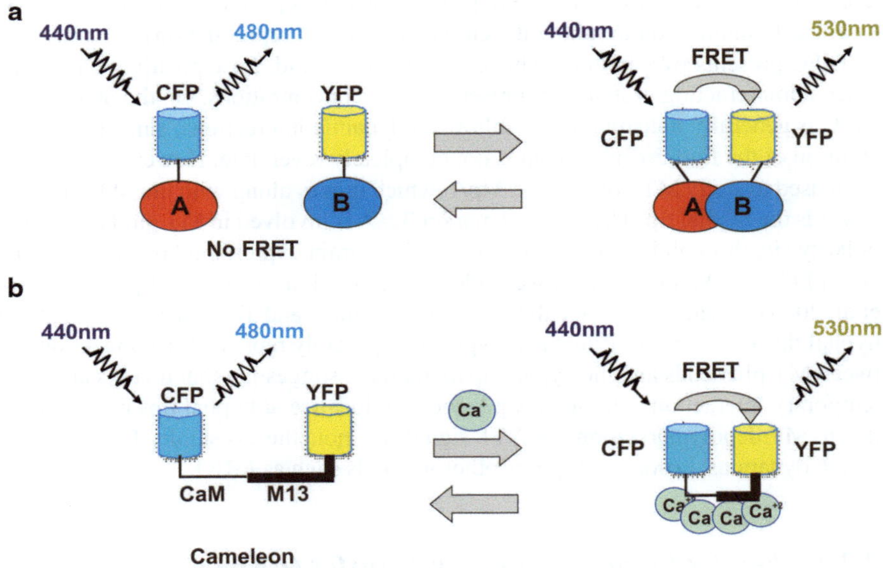

Fig. 2.2 Förster resonance energy transfer (FRET). **a** Schematic representation describing the FRET process between CFP- or YFP-tagged A and B proteins. If Protein A and Protein B do not interact, excitation at 440 nm by CFP leads to emission in cyan (480 nm). The fluorescence of the A protein could be followed in this channel. When Protein A and Protein B form a complex, excitation of Protein A with a 440-nm light is followed by an energy transfer (FRET) from CFP to YFP, which leads to a 530-nm yellow emission. **b** Schematic representation describing the background on the cameleon calcium biosensor. Cameleon consists of calmodulin (*CaM*) and the CaM-binding module M13 sandwiched between cyan (FRET donor) and yellow (acceptor) fluorescent proteins. In the absence of Ca^{2+}, excitation of CFP with a 440-nm light leads to the emission of cyan fluorescence. When the CaM module binds four Ca^{2+} ions and interacts with M13, a conformational change of CFP brings FRET to YFP, leading to the emission of yellow fluorescence. The interaction is reversible, thus allowing researchers to monitor Ca^{2+} concentration changes in living cells

of examples of FRET analysis in filamentous fungi is limited, FRET-based probes have the potential for growing impact in the fungal field, based on successful assays performed in other organisms to detect proteolytic activities, posttranslational modifications, enzymatic activities, Zn^{2+} or cyclic guanosine monophosphate (cGMP) (see references within Ibraheem and Campbell 2010).

2.2.3 *Fluorescence Lifetime Imaging Microscopy (FLIM) and FLIM–FRET*

FLIM bases contrast on the lifetime of individual fluorophores, rather than their emission spectra, producing an image based on the differences in the excited state decay rates. The fluorescence lifetime is defined as the average time that a molecule remains in an excited state prior to returning to the ground state by emitting a photon. As the fluorescence lifetime does not depend on concentration, absorption by the sample, sample thickness, photobleaching, and/or excitation intensity, FLIM is more convenient for quantification than intensity-based methods (Chang et al. 2007). It can also be applied to tracking signaling events inside living cells, such as monitoring changes in intracellular ions and detecting protein–protein interactions.

The principle behind FLIM–FRET relies on the fact that FRET is a very efficient fluorescence quencher and in the presence of a suitable acceptor the lifetime of the donor will decrease. Since the fluorescence decay of a donor/acceptor population contains the fluorescence of quenched and nonquenched molecules, quantitative FLIM–FRET measurements thus require the average lifetime of the donor with and without quenching so that FRET efficiency can be calculated based on the ratio thereof. FLIM–FRET analysis has been applied to the study of v-SNARE/t-SNARE interactions in *Trichoderma reesei* (Altenbach et al. 2009; Valkonen et al. 2007). The high spatial resolution imaging in living cells showed that the site of ternary SNARE complex formation is spatially segregated at distinct sites of the apical plasma membrane.

2.3 Protein Localization and Dynamics

This section will introduce advanced microscopy techniques, some of which are covered in depth in other chapters of this volume, that facilitate the analysis of protein transport and localization within the context of the cell cycle or the growth phase. Standardized fluorescent protein labeling methods for filamentous fungal proteins (Toews et al. 2004; Yang et al. 2004; Yu et al. 2004) in conjunction with fluorescence or confocal microscopy (see Chap. 1) have enabled *in vivo* analyses of protein dynamics. This has been achieved mainly through time-lapse acquirements over short- or long-time intervals, respectively, and superimposing or viewing the images as movies (see references below). Time-lapse fluorescence analysis of the proteins forming the nuclear pore complex (NPC), the multimeric channel that guides the transport of macromolecules between the nucleus and the cytoplasm (De Souza and Osmani 2007; Fiserova and Goldberg 2010) in *A. nidulans* showed that structural changes in the pore structure depend on the progression of the cell cycle (Osmani et al. 2006). Similarly, Markina-Iñarrairaegui and coworkers (Markina-Inarrairaegui et al. 2011) systematically analyzed the variable localization pattern of nuclear transporters, molecules involved in the translocation of cargoes through NPCs (Espeso and Osmani 2008), during the *A. nidulans* cell cycle (Fig. 2.3a). By streaming image data, the movement of several filamentous fungal proteins has been tracked. For example, anterograde and retrograde early-endosome (EE) transports between the tip and distal hyphal compartments have been shown in various filamentous fungal species to be facilitated by MTs (Fig. 2.3b; Abenza et al. 2009; Penalva et al. 2012; Steinberg 2012). Kinesin-3 and dynein motor proteins cooperate in this bidirectional transport of EEs (Schuster et al. 2011), which seems to be necessary for the transport of, for example, *A. nidulans* nuclear transporters such as karyopherins KapA and KapB (Etxebeste et al. 2013; Zekert and Fischer 2009).

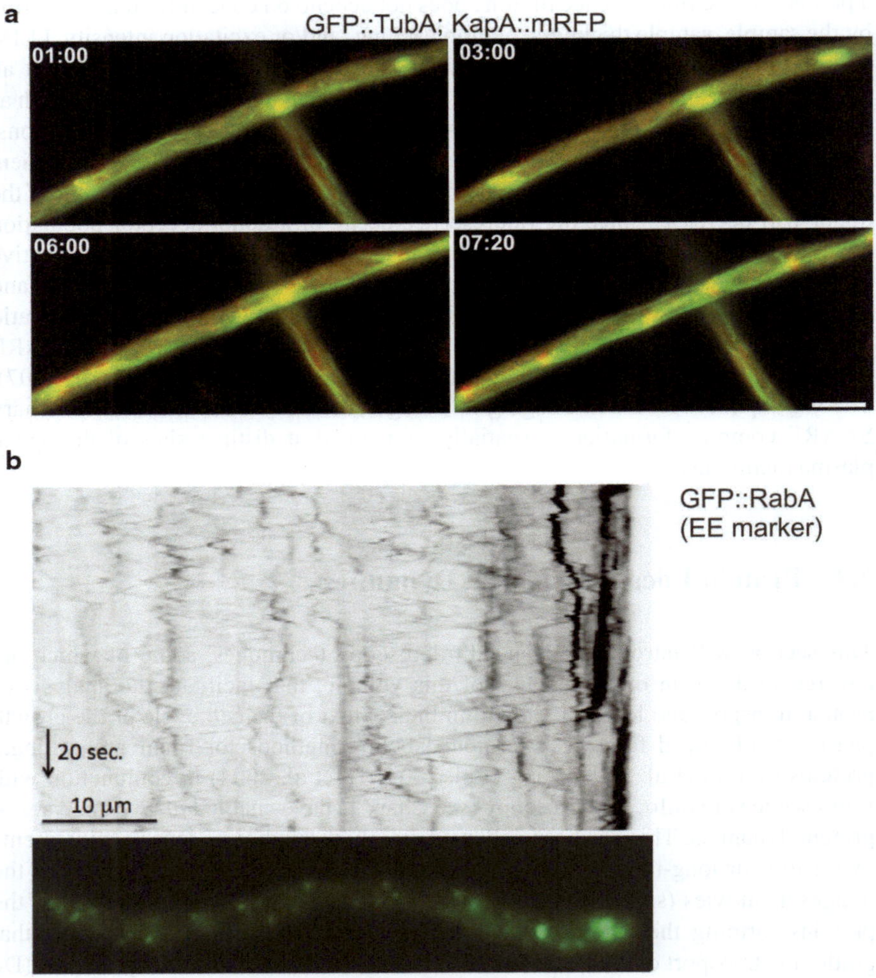

Fig. 2.3 Time-lapse and stream acquisition. **a** Selected frames from a time-lapse series (29 frames, 3 frames of 200 ms per min) showing a vegetative hypha of *Aspergillus nidulans* expressing KapA::mRFP and GFP::TubA chimeras during mitosis. *Yellow fluorescence* indicates co-localization of both proteins. Time is indicated in min:sec. Both proteins co-localize at the mitotic spindle. Scale bar = 5 μm. **b** Kymograph generated from a 20 s stream acquirement corresponding to a hypha expressing the early endosome marker RabA fused to GFP (Abenza et al. 2009). Scale bar = 10 μm

2.3.1 *Fluorescent Recovery After Photobleaching (FRAP)*

FRAP is another fluorescence-based microscopy technique which, when combined with time-lapse acquisitions and confocal microscopy (Chap. 1), can provide valuable information about the kinetic properties of a protein (Lippincott-Schwartz et al.

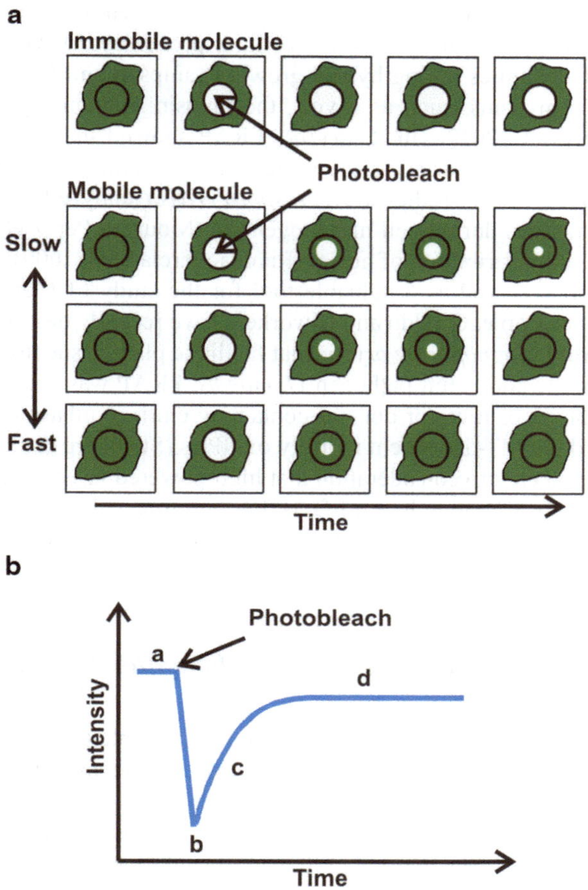

Fig. 2.4 Fluorescence recovery after photobleaching (FRAP). **a** Schematic representation showing the basis of the FRAP technique. Photobleaching of a region of interest (ROI) within the cell analyzed allows the measurement of the fluorescence recovery pace with molecules coming from the surrounding, non-photobleached, area. **b** Baseline of fluorescence *(a)* is reduced after photobleach *(b)*. Over time, the fluorescent intensity in the photobleached area increases as unbleached molecules diffuse into this area *(c)*. Later, there is a stabilization of the amount of fluorescence recovery *(d)*

2003). Photobleaching of a fluorophore, the extinction of its fluorescence through irradiation with directed high-intensity laser typically of short duration, is usually limited to a selected region (region of interest (ROI)) of the sample being analyzed (Fig. 2.4a; Klonis et al. 2002). After photobleaching, specific populations of fluorescently tagged proteins can be followed over time as they repopulate the ROI (Klonis et al. 2002; Lippincott-Schwartz et al. 2003). The rate of recovery is quantified to determine kinetics of protein movement, which can be fast, slow, or null, and can be mediated through binding/dissociation, diffusion, or transport processes (Fig. 2.4a, b; Lippincott-Schwartz et al. 2003).

Confocal (Chap. 1) FRAP is widely used and has been adapted to study a variety of filamentous fungal species and cellular processes. FRAP has been used to analyze MT dynamics during hyphal growth (Sampson and Heath 2005) or the role of specific myosins, such as MyoE, the myosin V homolog in *A. nidulans* (Taheri-Talesh et al., 2012). In *A. oryzae,* FRAP showed that secretory vesicles were constitutively transported to the septal plasma membrane (Hayakawa et al. 2011; Read 2011). FRAP studies in *Neurospora crassa* analyzed the MT polymerization and dynamics during cell morphogenesis (Mourino-Perez et al. 2006) and the movement and distribution of nuclei (Ramos-Garcia et al. 2009) .

FRAP has also served as a valuable tool for the study of filamentous fungal pathogens. For example, Giraldo and coworkers have recently described two secretion systems from *Magnaporthe oryzae* that facilitate plant tissue invasion (Giraldo et al. 2013). In the corn smut pathogen *U. maydis,* FRAP was used to analyze the dynamics of the myosin motor domain-containing chitin synthase Mcs1 (Schuster et al. 2012) and the MT-dependent motility of NPCs (Steinberg et al. 2012).

It should be taken into consideration that high-powered lasers could damage the sample being analyzed and induce artifacts. Furthermore, cell geometry and protein compartmentalization should be carefully studied for the selection of the appropriate model for the measurement of diffusion coefficients.

2.3.2 Photoconvertible Tagging of Filamentous Fungal Proteins

A recently developed array of FPs include photoswitchable (change between fluorescent and dark states), photoactivatable (fluorescence intensity increases upon irradiation), and photoconvertible fluorescent protein (PCFP; change from one emission wavelength to another) tags, whose spectral properties can be directly modified by a pulse of light (Chudakov et al. 2007b; Day and Davidson 2009). Dendra2 is a green-to-red monomeric PCFP which allows unique options for photolabeling and tracking of fusion proteins in living cells (Fig. 2.5a; Chudakov et al. 2007a, b; Gurskaya et al. 2006): (1) the tagged protein can be tracked in green before photoconversion; (2) Dendra2 can be photoactivated using either UV (405 nm) or blue (488 nm) light, and (3) the activated red Dendra2 signal renders high photostability (Chudakov et al. 2007b). Dendra2 has mostly been applied to studying higher eukaryotic cells and yeasts (see, e.g., references within Chudakov et al. 2007b; Jasik et al. 2013; Onischenko et al. 2009), but was recently used for the first time in filamentous fungi (Perez-de-Nanclares-Arregi and Etxebeste 2014).

The method was adapted to tracking low-expression proteins from *A. nidulans,* such as the apical developmental regulator FlbB (see Sect. 2.1; Etxebeste et al. 2009; Garzia et al. 2009; Harris 2009; Perez-de-Nanclares-Arregi and Etxebeste 2014), whose expression was driven through the constitutive *gpdA*^{mini} promoter (Pantazopoulou and Penalva 2009). As expected, green fluorescence was detected at the tip and nuclei before photoconversion, while red fluorescence was undetectable (Fig. 2.5b; Etxebeste et al. 2008; Etxebeste et al. 2009; Garzia et al. 2009).

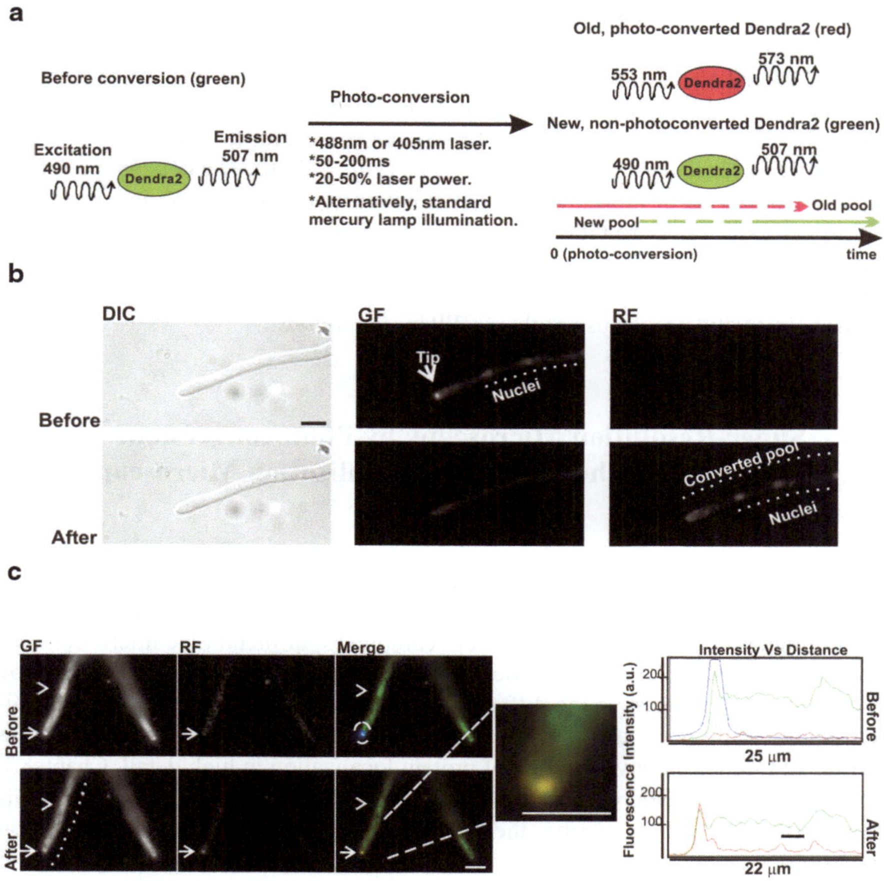

Fig. 2.5 The green-to-red photoconvertible protein Dendra2. **a** Before photoconversion, Dendra2 emits green fluorescence. Irradiation of a region of interest (ROI) with a 405-nm or 488-nm laser light or the entire cell with a standard mercury arc lamp provokes the photoconversion of Dendra2 to the red form. The old, photoconverted protein can be followed in *red*. New protein fluoresces in *green*. **b** DIC and *green (GF)* and *red (RF)* fluorescence images before and after irradiation with a mercury arc lamp of an *A. nidulans* hypha expressing FlbB::Dendra2 driven by the $gpdA^{mp}$ constitutive promoter. Light was filtered with the DAPI filter. **c** Selective photoconversion of the apical pool of FlbB::Dendra2. Both sets of panels show *green* and *red fluorescence* as well as merge images before and after photoconversion. Plots on the *right* represent fluorescence intensity along the *dotted line*. A magnification of the tip after photoconversion is also shown. Photoconversion of the tip was achieved by reducing the field of view with a diaphragm slider and exposing the ROI to UV light (80 % of the maximum intensity) during 2500 ms. *White arrows* and *arrowheads* indicate tips and the most apical nucleus, respectively. The *dotted circle* and the *blue color* represent the region irradiated with UV light. Scale bar in panels **b** and **c** = 5 µm. (Reproduced with permission from Perez-de-Nanclares-Arregi and Etxebeste 2014)

Photoconversion of Dendra2 was not induced through laser irradiation but rather using a UV-filtered standard mercury arc illumination of the entire field of view or an ROI (Fig. 2.5b, c, respectively; reproduced from Perez-de-Nanclares-Arregi and Etxebeste 2014), as previously described (Baker et al. 2010; Vorvis et al. 2008). After irradiation, green Dendra2 had been converted into the red form with significantly lower levels of green fluorescence (Fig. 2.5b, c). The dependence of the apical localization of FlbB was associated with active polar growth independently of the cell cycle (Herrero-Gracia E et al. 2015), supporting previous findings (Horio and Oakley 2005; Riquelme et al. 2003) showing that hyphal growth and cell cycle are independent processes in filamentous fungi (Riquelme et al. 2003). This proof of principle validates the use of this PCFP in *A. nidulans*.

2.4 Super-Resolution Microscopy by Photoconversion: mEosFP and Photoactivated Localization Microscopy (PALM)

In the past, protein transport analysis in filamentous fungal cells has been diffraction limited using conventional light microscopy, precluding accurate determination of nanometer-scale vesicles, protein complexes, or membrane micro-domains (membrane rafts, formerly identified as lipid rafts). Super-resolution microscopy techniques have overcome the diffraction limit (250 nm) of conventional light microscopy, resulting in lateral image resolution as high as 20 nm and providing a powerful tool for the investigation of protein localization in high detail. Chapter 3, "Super-Resolution Microscopy: SIM, STED and Localization Microscopy", written by Dodgson et al., describes the theory and application of the super-resolution microscopy techniques. Here we introduce one example of PALM using mEos in *A. nidulans.*

2.4.1 The Photoconvertible Fluorescent Protein mEosFP

EosFP, isolated from the Indopacific reef coral *Lobophyllia hemprichii* (Wiedenmann et al. 2004), is a tetrameric protein consisting of four identical protomers with a characteristic β-can fold of green fluorescent protein (GFP). A bright green fluorescent chromophore, generated spontaneously by autocatalysis, can be permanently photoconverted to red with near-UV irradiation. Protein engineering led to monomers of EosFP available and suitable for PALM, namely mEos2, mEosFP-*thermo,* and mIrisFP, the latter of which combines reversible on–off photoactivation switching with irreversible green-to-red photoconversion (McKinney et al. 2009; Wiedenmann et al. 2011). Both proteins have been used in a variety of applications that demonstrate the versatility of these marker proteins in live-cell imaging (Wiedenmann et al. 2011).

2.4.2 Photoactivated Localization Microscopy (PALM) Analysis in A. nidulans

As discussed in Chap. 3, PALM uses photoswitchable fluorophores to achieve temporal control of the emission through conversion between fluorescent ("on") and dark ("off") states. When sample excitation is of a sufficiently low intensity, only a random, sparse fluorophore subset will be in the "on" state at any time, allowing these molecules to be imaged individually, precisely localized, and then deactivated by switching to a reversible dark state or permanent bleaching. Such a strategy leads to the construction of a super-resolution image.

The *A. nidulans* apical membrane-associated polarity marker TeaR (see Sect. 2.1; Takeshita et al. 2008) was fused with mEosFP*thermo* and its localization visualized by PALM (Ishitsuka et al., in press). The size of TeaR domains at the apical membrane imaged by widefield epifluorescent microscopy was on the order of 440 nm (Fig. 2.6a upper picture, b green line), whereas PALM revealed the size to be closer to 100 nm (Fig. 2.6a lower picture, b red line). Since PALM analysis treats single molecules, it is suitable for quantified analysis. To put TeaR clusters in the context of membrane domains, four domains were identified in the PALM image (Fig. 2.6a, 2.6d) using cluster analysis (threshold = 10 neighboring molecules, within distance r = 50 nm) (Fig. 2.6c) . The number of TeaR molecules identified within each domain and the size of the domains are shown in Fig. 2.6e. It was also possible to track the trajectory of single TeaR molecules for several seconds prior to their photobleaching (shown with arrows in Fig. 2.6f; different colors indicate different duration). The movement of TeaR molecules was usually restricted to the membrane domains.

2.5 Conclusions and Future Perspectives

Microscopy techniques are being continuously developed and are becoming more accessible. A variety of techniques allow the analysis of protein localization, interactions, and dynamic behavior. When planning experiments, consideration must be given to the advantages and limitations of each method and its suitability to each specific scientific question.

While labeling filamentous fungal proteins with FPs is standard within the fungal community for protein localization, time-lapse and high-speed imaging of these chimeras is not always feasible based on instrument availability. The BiFC technique has been firmly established and widely used to confirm interactions between different protein pairs, but is problematic for determining dynamic interactions based on the irreversibility of the fluorescent protein reconstitution. Even if FRET analysis does not yield a positive signal between two proteins, a hypothetic interaction cannot be excluded based on false negatives. Photoconvertible FPs such as Dendra2 provide an attractive tool to follow protein dynamics, but suffer from limited brightness. A hypothetical impact of UV or high-powered light on cells must

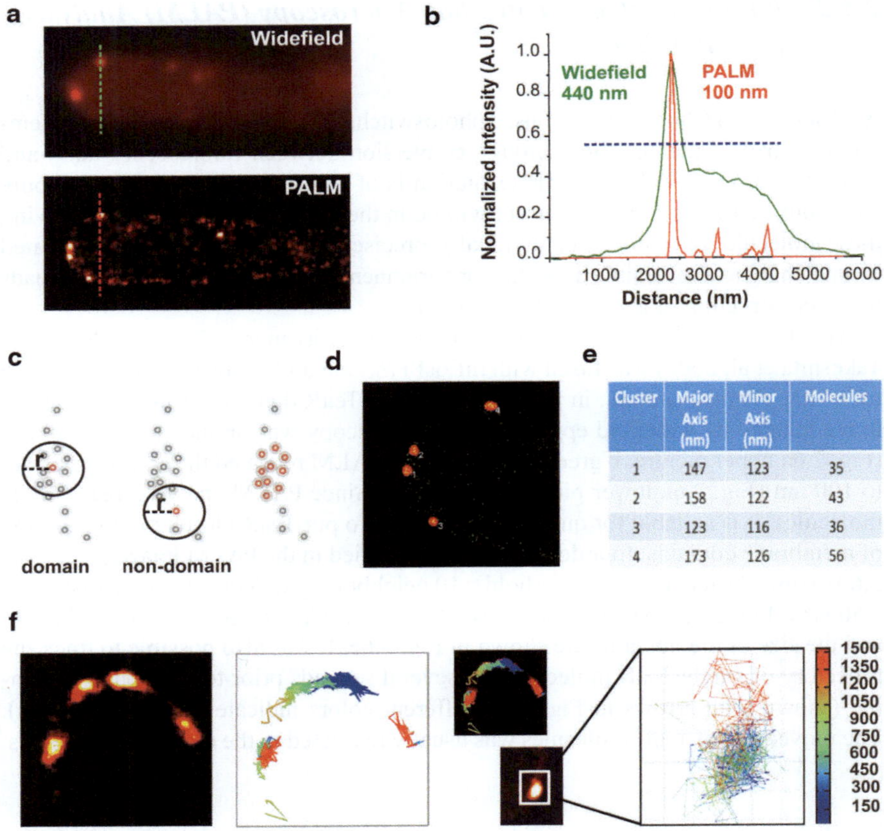

Cluster	Major Axis (nm)	Minor Axis (nm)	Molecules
1	147	123	35
2	158	122	43
3	123	116	36
4	173	126	56

Fig. 2.6 PALM analysis of the membrane-associated polarity marker TeaR in *A. nidulans*. **a**, **b** The size of TeaR domains imaged by widefield epifluorescent microscopy is approximately 440 nm *(upper, green line)*, whereas PALM revealed that the size is actually closer to 100 nm (*lower, red line*, PALM image is constructed by 1000 frames, time resolution 50 ms). **c** The membrane domains were defined by cluster analysis, setting the threshold on the number of neighboring molecules (ten molecules) within a certain distance r (50 nm). **d** Four domains were defined by the cluster analysis on the PALM image of panel **a**. **e** The size of domains and number of molecules within the domains are shown. **f** The dynamic behavior of single molecule can be traced for a few seconds by PALM, until they are photobleached. Each *line* indicates the movement of a TeaR single molecule. Different colors indicate different timing. The movement is usually restricted within the membrane domains

also be considered in FRAP or photoconversion experiments. Taking everything into consideration, it is important to select the microscopy technique that best fits the aim of the study.

Super-resolution microscopy techniques, a more recent development that breaks the diffraction barrier, require specialized and expensive instrumentation, but, in combination with molecular techniques, can provide valuable information on protein localization and dynamics in the small fungal cell. Specific and efficient delivery of bright fluorophores into the target structure without adding substantial background is required for several single-molecule nanoscopy approaches. Bright organic dyes can be delivered via antibodies. Nanobodies are a unique form of monoclonal antibodies, characterized by a single antigen-binding domain (12–15 kDa) present in camels (Hamers-Casterman et al. 1993). The nanobody can be easily produced in bacteria or yeast and binds to the antigen with high affinity and specificity. The diverse application of nanobodies will be important in general cell biology (Ries et al. 2012).

Overall, the advanced microscopy techniques reviewed in this chapter open new avenues for experimentation that promise to generate key information on the molecular mechanisms that control signal transduction, morphogenesis, and development of filamentous fungi.

Acknowledgments Work at N. Takeshita's laboratory was funded by the DFG (TA819/2) and the Baden-Württemberg Stiftung. Work by O. Etxebeste was funded by the Basque Government through grant IT599-13 and the Ministerio de Economía y Competitividad (formerly Ministerio de Ciencia e Innovación) through grant BFU2010-17528. We would like to express our gratitude to Dr. Marc S. Cortese for proofreading the document.

References

Abenza JF, Pantazopoulou A, Rodriguez JM, Galindo A, Penalva MA (2009) Long-distance movement of *Aspergillus nidulans* early endosomes on microtubule tracks. Traffic 10(1):57–75. (Available from PM: 19000168)

Adams TH, Wieser JK, Yu J-H (1998) Asexual sporulation in *Aspergillus nidulans*. Microbiol Mol Biol Rev 62(1):35–54. http://mmbr.asm.org/cgi/content/abstract/62/1/35 (Available from PM: 9529886)

Allen GJ, Kwak JM, Chu SP, Llopis J, Tsien RY, Harper JF, Schroeder JI (1999) Cameleon calcium indicator reports cytoplasmic calcium dynamics in *Arabidopsis* guard cells. Plant J 19(6):735–747. (Available from PM: 10571859)

Altenbach K, Duncan RR, Valkonen M (2009) In vivo FLIM-FRET measurements of recombinant proteins expressed in filamentous fungi. Fungal Biol Rev 23(3):67–71. http://www.sciencedirect.com/science/article/pii/S1749461310000023 doi:10.1016/j.fbr.2009.12.002

Baker SM, Buckheit RW III, Falk MM (2010) Green-to-red photoconvertible fluorescent proteins: tracking cell and protein dynamics on standard wide-field mercury arc-based microscopes. BMC Cell Biol 11:15. (Available from PM: 20175925)

Bayram O, Braus GH, Fischer R, Rodriguez-Romero J (2010) Spotlight on *Aspergillus nidulans* photosensory systems. Fungal Genet Biol 47(11):900–908. (Available from PM: 20573560)

Brand A (2012) Hyphal growth in human fungal pathogens and its role in virulence. Int J Microbiol 2012:517529. (Available from PM: 22121367)

Chang CW, Sud D, Mycek MA (2007) Fluorescence lifetime imaging microscopy. Methods Cell Biol 81:495–524. (Available from PM: 17519182)

Chudakov DM, Lukyanov S, Lukyanov KA (2007a) Tracking intracellular protein movements using photoswitchable fluorescent proteins PS-CFP2 and Dendra2. Nat Protoc 2(8):2024–2032. (Available from PM: 17703215)

Chudakov DM, Lukyanov S, Lukyanov KA (2007b) Using photoactivatable fluorescent protein Dendra2 to track protein movement. Biotechniques 42(5)553, 555, 557. (Available from PM: 17515192)

Day RN, Davidson MW (2009) The fluorescent protein palette: tools for cellular imaging. Chem Soc Rev 38(10):2887–2921. (Available from PM: 19771335)

De Souza CP, Osmani SA (2007) Mitosis, not just open or closed. Eukaryot Cell 6(9):1521–1527. (Available from PM: 17660363)

Espeso EA, Osmani SA (2008) Nuclear pore complex and transport in *Aspergillus nidulans*. In: Goldman GH, Osmani SA (eds) The *Aspergilli*: genomics, medical aspects, biotechnology, and research methods. Taylor & Francis Group, Boca Raton, pp 261–277

Etxebeste O, Ni M, Garzia A, Kwon NJ, Fischer R, Yu JH, Espeso EA, Ugalde U (2008) Basic-zipper-type transcription factor FlbB controls asexual development in *Aspergillus nidulans*. Eukaryot Cell 7(1):38–48. (Available from PM: 17993569)

Etxebeste O, Herrero-Garcia E, Araujo-Bazan L, Rodriguez-Urra AB, Garzia A, Ugalde U, Espeso EA (2009) The bZIP-type transcription factor FlbB regulates distinct morphogenetic stages of colony formation in *Aspergillus nidulans*. Mol Microbiol 73(5):775–789. (Available from PM: 19656299)

Etxebeste O, Ugalde U, Espeso EA (2010) Adaptative and developmental responses to stress in *Aspergillus nidulans*. Curr Protein Pept Sci 11(8):704–718. (Available from PM: 21235506)

Etxebeste O, Villarino M, Markina-Inarrairaegui A, Araujo-Bazan L, Espeso EA (2013) Cytoplasmic dynamics of the general nuclear import machinery in apically growing syncytial cells. PLoS ONE 8(12):e85076. (Available from PM: 24376868)

Fischer R, Timberlake WE (1995) *Aspergillus nidulans* apsA (anucleate primary sterigmata) encodes a coiled-coil protein required for nuclear positioning and completion of asexual development. J Cell Biol 128(4):485–498. (Available from PM: 7860626)

Fischer R, Zekert N, Takeshita N (2008) Polarized growth in fungi—interplay between the cytoskeleton, positional markers and membrane domains. Mol Microbiol 68(4):813–826. (Available from PM: 18399939)

Fiserova J, Goldberg MW (2010) Nucleocytoplasmic transport in yeast: a few roles for many actors. Biochem Soc Trans 38(Pt 1):273–277. (Available from PM: 20074073)

Fisher MC, Henk DA, Briggs CJ, Brownstein JS, Madoff LC, McCraw SL, Gurr SJ (2012) Emerging fungal threats to animal, plant and ecosystem health. Nature 484(7393):186–194. (Available from PM: 22498624)

Garzia A, Etxebeste O, Herrero-Garcia E, Fischer R, Espeso EA, Ugalde U (2009) *Aspergillus nidulans* FlbE is an upstream developmental activator of conidiation functionally associated with the putative transcription factor FlbB. Mol Microbiol 71(1):172–184. (Available from PM: 19007409)

Garzia A, Etxebeste O, Herrero-Garcia E, Ugalde U, Espeso EA (2010) The concerted action of bZip and cMyb transcription factors FlbB and FlbD induces brlA expression and asexual development in *Aspergillus nidulans*. Mol Microbiol 75(5):1314–1324. (Available from PM: 20132447)

Gibbs PA, Seviour RJ, Schmid F (2000) Growth of filamentous fungi in submerged culture: problems and possible solutions. Crit Rev Biotechnol 20(1):17–48. (Available from PM: 10770226)

Giraldo MC, Dagdas YF, Gupta YK, Mentlak TA, Yi M, Martinez-Rocha AL, Saitoh H, Terauchi R, Talbot NJ, Valent B (2013) Two distinct secretion systems facilitate tissue invasion by the rice blast fungus *Magnaporthe oryzae*. Nat Commun 4:1996. (Available from PM: 23774898)

Gregory PH (1966) The fungus spore: what it is and what it does. In: Madelin MF (ed) The fungus spore. Butterworths, London, pp 1–13

Gurskaya NG, Verkhusha VV, Shcheglov AS, Staroverov DB, Chepurnykh TV, Fradkov AF, Lukyanov S, Lukyanov KA (2006) Engineering of a monomeric green-to-red photoactivatable fluorescent protein induced by blue light. Nat Biotechnol 24(4):461–465. (Available from PM: 16550175)

Gutierrez-Correa M, Ludena Y, Ramage G, Villena GK (2012) Recent advances on filamentous fungal biofilms for industrial uses. Appl Biochem Biotechnol 167(5):1235–1253. (Available from PM: 22350934)

Hamers-Casterman C, Atarhouch T, Muyldermans S, Robinson G, Hamers C, Songa EB, Bendahman N, Hamers R (1993) Naturally occurring antibodies devoid of light chains. Nature 363(6428):446–448. (Available from PM: 8502296)

Harris SD (2006) Cell polarity in filamentous fungi: shaping the mold. Int Rev Cytol 251:41–77. (Available from PM: 16939777)

Harris SD (2009) The Spitzenkorper: a signalling hub for the control of fungal development? Mol Microbiol 73(5):733–736. (Available from PM: 19627503)

Harris SD, Morrell JL, Hamer JE (1994) Identification and characterization of *Aspergillus nidulans* mutants defective in cytokinesis. Genetics 136(2):517–532. (Available from PM: 8150280)

Hayakawa Y, Ishikawa E, Shoji JY, Nakano H, Kitamoto K (2011) Septum-directed secretion in the filamentous fungus *Aspergillus oryzae*. Mol Microbiol 81(1):40–55. (Available from PM: 21564341)

Herrero-Garcia E, Perez-de-Nanclares-Arregi E, Cortese MS, Markina-Iñarrairaegui A, Oiartzabal-Arano E, Etxebeste O, Ugalde U, Espeso EA (2015) Tip-to-nucleus migration dynamics of the asexual development regulator FlbB in vegetative cells. Mol Microbiol. doi: 10.1111/mmi.13156.

Higashitsuji Y, Herrero S, Takeshita N, Fischer R (2009) The cell end marker protein TeaC is involved in growth directionality and septation in *Aspergillus nidulans*. Eukaryot Cell 8(7):957–967. (Available from PM: 19429780)

Horio T, Oakley BR (2005) The role of microtubules in rapid hyphal tip growth of *Aspergillus nidulans*. Mol Biol Cell 16(2):918–926. (Available from PM: 15548594)

Ibraheem A, Campbell RE (2010) Designs and applications of fluorescent protein-based biosensors. Curr Opin Chem Biol 14(1):30–36. (Available from PM: 19913453)

Ishitsuka Y, Savage N, Li Y, Bergs A, Gruen N, Kohler D, Donnelly R, Nienhaus GU, Fischer R, Takeshita N. (2015) Super-resolution microscopy reveals a dynamic picture of cell polarity maintenance during directional growth. Sci Adv, In press

Jasik J, Boggetti B, Baluska F, Volkmann D, Gensch T, Rutten T, Altmann T, Schmelzer E (2013) PIN2 turnover in *Arabidopsis* root epidermal cells explored by the photoconvertible protein Dendra2. PLoS ONE 8(4):e61403. (Available from PM: 23637828)

Jiang D, Zhu W, Wang Y, Sun C, Zhang KQ, Yang J (2013) Molecular tools for functional genomics in filamentous fungi: recent advances and new strategies. Biotechnol Adv 31(8):1562–1574. (Available from PM: 23988676)

Kerppola TK (2006) Design and implementation of bimolecular fluorescence complementation (BiFC) assays for the visualization of protein interactions in living cells. Nat Protoc 1(3):1278–1286. (Available from PM: 17406412)

Kim HS, Czymmek KJ, Patel A, Modla S, Nohe A, Duncan R, Gilroy S, Kang S (2012) Expression of the cameleon calcium biosensor in fungi reveals distinct Ca^{2+} signatures associated with polarized growth, development, and pathogenesis. Fungal Genet Biol 49(8):589–601. (Available from PM: 22683653)

Klonis N, Rug M, Harper I, Wickham M, Cowman A, Tilley L (2002) Fluorescence photobleaching analysis for the study of cellular dynamics. Eur Biophys J 31(1):36–51. (Available from PM: 12046896)

Kodama Y, Hu CD (2012) Bimolecular fluorescence complementation (BiFC): a 5-year update and future perspectives. Biotechniques 53(5):285–298. (Available from PM: 23148879)

Kües U, Fischer R (2006) Asexual sporulation in mycelial fungi. In: Kües U, Fischer R (eds) The Mycota vol. I: growth, differentiation and sexuality. Springer-Verlag, Berlin, pp 263–292

Kürnsteiner H, Zinner M, Kück U (2002) Immunosuppressants. In: Esser K, Bennett JW (eds) The Mycota X: industrial application. Springer-Verlag, Berlin, pp 129–155

Lippincott-Schwartz J, Altan-Bonnet N, Patterson GH (2003) Photobleaching and photoactivation: following protein dynamics in living cells. Nat Cell Biol 5:S7–S14. (Available from PM: 14562845)

Markina-Inarrairaegui A, Etxebeste O, Herrero-Garcia E, Araujo-Bazan L, Fernandez-Martinez J, Flores JA, Osmani SA, Espeso EA (2011) Nuclear transporters in a multinucleated organism:

functional and localization analyses in *Aspergillus nidulans*. Mol Biol Cell 22(20):3874–3886. (Available from PM: 21880896)

McKinney SA, Murphy CS, Hazelwood KL, Davidson MW, Looger LL (2009) A bright and photostable photoconvertible fluorescent protein. Nat Methods 6(2):131–133. (Available from PM: 19169260)

Momany M (2002) Polarity in filamentous fungi: establishment, maintenance and new axes. Curr Opin Microbiol 5(6):580–585. (Available from PM: 12457701)

Momany M, Hamer JE (1997) Relationship of actin, microtubules, and crosswall synthesis during septation in *Aspergillus nidulans*. Cell Motil Cytoskeleton 38(4):373–384. (Available from PM: 9415379)

Mourino-Perez RR, Roberson RW, Bartnicki-Garcia S (2006) Microtubule dynamics and organization during hyphal growth and branching in *Neurospora crassa*. Fungal Genet Biol 43(6):389–400. (Available from PM: 16621627)

Nagai T, Yamada S, Tominaga T, Ichikawa M, Miyawaki A (2004) Expanded dynamic range of fluorescent indicators for Ca^{2+} by circularly permuted yellow fluorescent proteins. Proc Natl Acad Sci U S A 101(29):10554–10559. (Available from PM: 15247428)

Onischenko E, Stanton LH, Madrid AS, Kieselbach T, Weis K (2009) Role of the Ndc1 interaction network in yeast nuclear pore complex assembly and maintenance. J Cell Biol 185(3):475–491. (Available from PM: 19414609)

Osmani AH, Davies J, Liu HL, Nile A, Osmani SA (2006) Systematic deletion and mitotic localization of the nuclear pore complex proteins of *Aspergillus nidulans*. Mol Biol Cell 17(12):4946–4961. (Available from PM: 16987955)

Pantazopoulou A, Penalva MA (2009) Organization and dynamics of the *Aspergillus nidulans* golgi during apical extension and mitosis. Mol Biol Cell 20(20):4335–4347. (Available from PM: 19692566)

Penalva MA, Galindo A, Abenza JF, Pinar M, Calcagno-Pizarelli AM, Arst HN, Pantazopoulou A (2012) Searching for gold beyond mitosis: mining intracellular membrane traffic in *Aspergillus nidulans*. Cell Logist 2(1):2–14. (Available from PM: 22645705)

Perez-de-Nanclares-Arregi E, Etxebeste O (2014) Photo-convertible tagging for localization and dynamic analyses of low-expression proteins in filamentous fungi. Fungal Genet Biol 70:33–41. (Available from PM: 25014896)

Pontecorvo G, Roper JA, Hemmons LM, McDonald KD, Bufton AW (1953) The genetics of *Aspergillus nidulans*. Adv Genet 5:141–238

Purschwitz J, Muller S, Kastner C, Schoser M, Haas H, Espeso EA, Atoui A, Calvo AM, Fischer R (2008) Functional and physical interaction of blue- and red-light sensors in *Aspergillus nidulans*. Curr Biol 18(4):255–259. (Available from PM: 18291652)

Raffaele S, Kamoun S (2012) Genome evolution in filamentous plant pathogens: why bigger can be better. Nat Rev Microbiol 10(6):417–430. http://dx.doi.org/10.1038/nrmicro2790 (Available from PM: 22565130)

Ramos-Garcia SL, Roberson RW, Freitag M, Bartnicki-Garcia S, Mourino-Perez RR (2009) Cytoplasmic bulk flow propels nuclei in mature hyphae of *Neurospora crassa*. Eukaryot Cell 8(12):1880–1890. (Available from PM: 19684281)

Read ND (2011) Exocytosis and growth do not occur only at hyphal tips. Mol Microbiol 81(1):4–7. (Available from PM: 21645129)

Ries J, Kaplan C, Platonova E, Eghlidi H, Ewers H (2012) A simple, versatile method for GFP-based super-resolution microscopy via nanobodies. Nat Methods 9(6):582–584. (Available from PM: 22543348)

Riquelme M (2013) Tip growth in filamentous fungi: a road trip to the apex. Annu Rev Microbiol 67:587–609. (Available from PM: 23808332)

Riquelme M, Fischer R, Bartnicki-Garcia S (2003) Apical growth and mitosis are independent processes in *Aspergillus nidulans*. Protoplasma 222(3–4):211–215. (Available from PM: 14714210)

Rodriguez-Romero J, Hedtke M, Kastner C, Muller S, Fischer R (2010) Fungi, hidden in soil or up in the air: light makes a difference. Annu Rev Microbiol 64:585–610. (Available from PM: 20533875)

Rogers KL, Picaud S, Roncali E, Boisgard R, Colasante C, Stinnakre J, Tavitian B, Brulet P (2007) Non-invasive in vivo imaging of calcium signaling in mice. PLoS ONE 2(10):e974. (Available from PM: 17912353)

Sampson K, Heath IB (2005) The dynamic behaviour of microtubules and their contributions to hyphal tip growth in *Aspergillus nidulans*. Microbiology 151(Pt 5):1543–1555. (Available from PM: 15870464)

Schuster M, Kilaru S, Fink G, Collemare J, Roger Y, Steinberg G (2011) Kinesin-3 and dynein cooperate in long-range retrograde endosome motility along a nonuniform microtubule array. Mol Biol Cell 22(19):3645–3657. (Available from PM: 21832152)

Schuster M, Treitschke S, Kilaru S, Molloy J, Harmer NJ, Steinberg G (2012) Myosin-5, kinesin-1 and myosin-17 cooperate in secretion of fungal chitin synthase. EMBO J 31(1):214–227. (Available from PM: 22027862)

Sekar RB, Periasamy A (2003) Fluorescence resonance energy transfer (FRET) microscopy imaging of live cell protein localizations. J Cell Biol 160(5):629–633. (Available from PM: 12615908)

Soanes DM, Alam I, Cornell M, Wong HM, Hedeler C, Paton NW, Rattray M, Hubbard SJ, Oliver SG, Talbot NJ (2008) Comparative genome analysis of filamentous fungi reveals gene family expansions associated with fungal pathogenesis. PLoS ONE 3(6):e2300. (Available from PM: 18523684)

Steinberg G (2012) The transport machinery for motility of fungal endosomes. Fungal Genet Biol 49(9):675–676. (Available from PM: 22330672)

Steinberg G, Schuster M, Theisen U, Kilaru S, Forge A, Martin-Urdiroz M (2012) Motor-driven motility of fungal nuclear pores organizes chromosomes and fosters nucleocytoplasmic transport. J Cell Biol 198(3):343–355. (Available from PM: 22851316)

Taheri-Talesh N, Xiong Y, Oakley BR (2012) The functions of myosin II and myosin V homologs in tip growth and septation in *Aspergillus nidulans*. PLoS ONE 7(2):e31218. (Available from PM: 22359575)

Takeshita N, Higashitsuji Y, Konzack S, Fischer R (2008) Apical sterol-rich membranes are essential for localizing cell end markers that determine growth directionality in the filamentous fungus *Aspergillus nidulans*. Mol Biol Cell 19(1):339–351. (Available from PM: 18003978)

Takeshita N, Mania D, Herrero S, Ishitsuka Y, Nienhaus GU, Podolski M, Howard J, Fischer R (2013) The cell-end marker TeaA and the microtubule polymerase AlpA contribute to microtubule guidance at the hyphal tip cortex of *Aspergillus nidulans* to provide polarity maintenance. J Cell Sci 126(Pt 23):5400–5411. (Available from PM: 24101725)

Takeshita N, Manck R, Grun N, de Vega SH, Fischer R (2014) Interdependence of the actin and the microtubule cytoskeleton during fungal growth. Curr Opin Microbiol 20:34–41. (Available from PM: 24879477)

Todd RB, Davis MA, Hynes MJ (2007) Genetic manipulation of *Aspergillus nidulans*: meiotic progeny for genetic analysis and strain construction. Nat Protoc 2(4):811–821. http://dx.doi.org/10.1038/nprot.2007.112 (Available from PM: 17446881)

Toews MW, Warmbold J, Konzack S, Rischitor P, Veith D, Vienken K, Vinuesa C, Wei H, Fischer R (2004) Establishment of mRFP1 as a fluorescent marker in *Aspergillus nidulans* and construction of expression vectors for high-throughput protein tagging using recombination in vitro (GATEWAY). Curr Genet 45(6):383–389. (Available from PM: 15071756)

Tsien RY (1998) The green fluorescent protein. Annu Rev Biochem 67:509–544. (Available from PM: 9759496)

Valkonen M, Kalkman ER, Saloheimo M, Penttila M, Read ND, Duncan RR (2007) Spatially segregated SNARE protein interactions in living fungal cells. J Biol Chem 282(31):22775–22785. (Available from PM: 17553800)

Vorvis C, Markus SM, Lee WL (2008) Photoactivatable GFP tagging cassettes for protein-tracking studies in the budding yeast *Saccharomyces cerevisiae*. Yeast 25(9):651–659. (Available from PM: 18727145)

Wiedenmann J, Ivanchenko S, Oswald F, Schmitt F, Rocker C, Salih A, Spindler KD, Nienhaus GU (2004) EosFP, a fluorescent marker protein with UV-inducible green-to-red fluorescence conversion. Proc Natl Acad Sci U S A 101(45):15905–15910. (Available from PM: 15505211)
Wiedenmann J, Gayda S, Adam V, Oswald F, Nienhaus K, Bourgeois D, Nienhaus GU (2011) From EosFP to mIrisFP: structure-based development of advanced photoactivatable marker proteins of the GFP-family. J Biophotonics 4(6):377–390. (Available from PM: 21319305)
Yang L, Ukil L, Osmani A, Nahm F, Davies J, De Souza CP, Dou X, Perez-Balaguer A, Osmani SA (2004) Rapid production of gene replacement constructs and generation of a green fluorescent protein-tagged centromeric marker in *Aspergillus nidulans*. Eukaryot Cell 3(5):1359–1362. (Available from PM: 15470263)
Yu JH, Hamari Z, Han KH, Seo JA, Reyes-Dominguez Y, Scazzocchio C (2004) Double-joint PCR: a PCR-based molecular tool for gene manipulations in filamentous fungi. Fungal Genet Biol 41(11):973–981. (Available from PM: 15465386)
Zekert N, Fischer R (2009) The *Aspergillus nidulans* kinesin-3 UncA motor moves vesicles along a subpopulation of microtubules. Mol Biol Cell 20(2):673–684. (Available from PM: 19037104)

Chapter 3
Super-Resolution Microscopy: SIM, STED and Localization Microscopy

James Dodgson, Anatole Chessel, Susan Cox and Rafael E. Carazo Salas

3.1 Introduction

Before looking at the individual super-resolution techniques, it is necessary to briefly introduce or revise a few key concepts that will facilitate understanding them.

The Microscope Is a Low-Pass Spatial Frequency Filter In very simple terms, a microscope illuminates a sample, collects the light diffracted by the sample with a lens and then builds an interference pattern that is viewed as an image on the back focal plane. Not all the diffracted light is collected by the lens (Fig. 3.1a), and consequently the highest spatial information is lost. A diffraction-limited image is therefore "fuzzy" or "blurred". This is demonstrated in Fig. 3.1b, where the point spread function (PSF) of a single point source of light (a single fluorescent bead 100 nm in diameter) is shown. In "conventional" or "diffraction-limited" microscopy, the PSF will be of a size that depends on the precise optics and on the wavelength of light used. For fluorescence microscopy, this would not generally be smaller than ~200 nm in the lateral plane and ~500 nm in the axial plane. This is the resolution

---

J. Dodgson (✉) · R. E. Carazo Salas
Genetics Department, University of Cambridge, Downing Street, Cambridge
CB2 1EH, UK
e-mail: jd535@cam.ac.uk

R. E. Carazo Salas
e-mail: cre20@cam.ac.uk

A. Chessel
Department of Genetics, University of Cambridge, Tennis Court Road, Cambridge
CB2, 1PD, UK
e-mail: ac744@cam.ac.uk

S. Cox
Randall Division of Cell and Molecular Biophysics, King's College London,
New Hunt's House, Guy's Campus, London SE1 1UL, UK
e-mail: susan.cox@kcl.ac.uk

T. E. S. Dahms and K. J. Czymmek (eds.), *Advanced Microscopy in Mycology*,
Fungal Biology, DOI 10.1007/978-3-319-22437-4_3

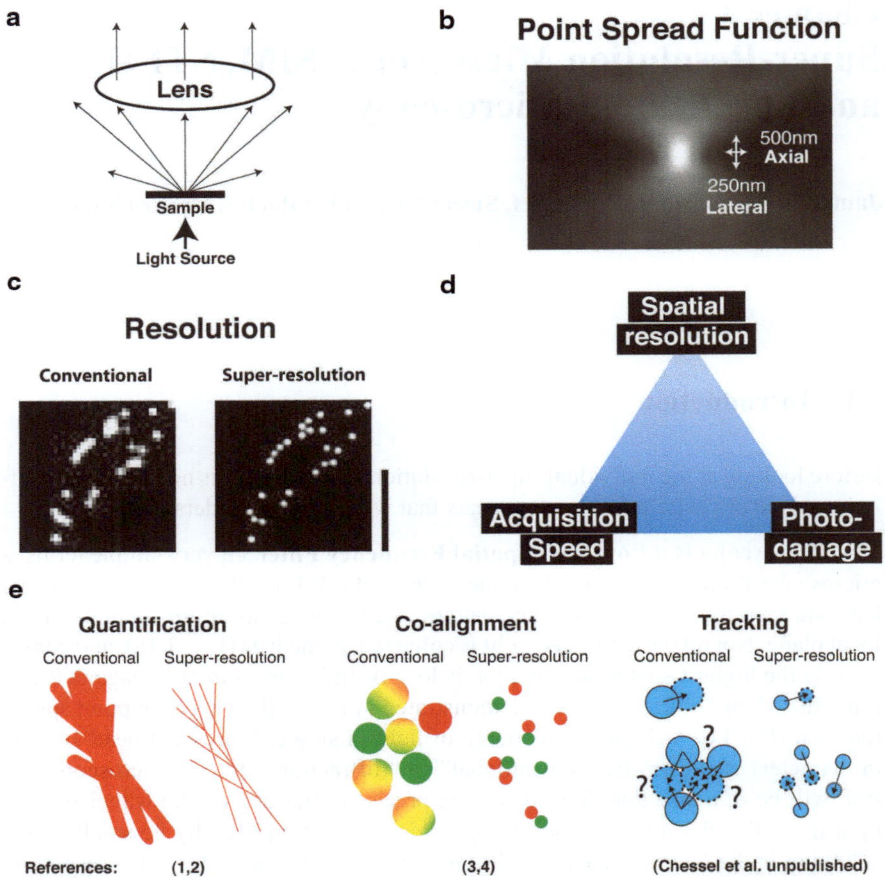

Fig. 3.1 Basic concepts. **a** The lens fails to collect all the diffracted light from a sample being illuminated, and as a consequence high-resolution spatial information is lost. **b** The point spread function of a 100-nm-diameter fluorescent glass bead is 250 nm laterally and 500 nm axially. **c** Conventional imaging cannot allow many of the proximal glass beads to be distinguished, but super-resolution imaging can. **d** The super-resolution imaging-limitation triangle (as formulated by and with kind permission from Eric Betzig). **e** Graphical representation of the some of the potential insights that super-resolution microscopy has to provide to fungal cell imaging. Scale bar is 2 μm

limit, and it restricts the potential of imaging to resolve (or distinguish between) two points that are close together (Fig. 3.1c).

Super-Resolution Super-resolution breaks the diffraction limit, allowing the differentiation of conventionally unresolvable objects (see Fig. 3.1c). This relatively recent development is reviewed in the context of three main super-resolution techniques currently available. When evaluating the relative merits of each technique, we need to consider the limitations of spatial resolution (both lateral and axial), photodamage (a combination of photobleaching and phototoxicity) and acquisition speed. Improvements in any one factor are usually to the detriment of the other two, expressed in the form of "the super-resolution imaging-limitation triangle" (Fig. 3.1d). These limitations are especially crucial in the realm of live-cell imaging. Thus the first, and maybe most important, part of any super-resolution experiment is to choose the most appropriate technique.

In the following sections, we give a primer of three super-resolution imaging techniques as applied to mycological research. As the field of super-resolution microscopy is one that is wide and rapidly evolving, rather than doing a comprehensive review, we stay within the confines of the more mainstream techniques that have been commercialized and demonstrated to be useful for work in fungi.

The Value of Super-Resolution Imaging The value of these techniques to fungal cell biology is demonstrated by the citation of specific examples in fungal live- and fixed-cell imaging. Figure panel 3.1e provides a graphical summary of some of the potential insights that super-resolution imaging can unlock, including the accurate quantification of structures including their number and size (Miao et al. 2013; Puchner et al. 2013), an accurate assessment of co-alignment and other spatial arrangements between structures (Daboussi et al. 2012; Voelkel-Meiman et al. 2013) and allowing the tracking of dynamic structures otherwise too densely packed to distinguish (Chessel et al. Unpublished). This is by no means a comprehensive list, and many other applications are and will be possible for techniques that reveal the cell in such high levels of detail.

3.2 Structured Illumination Microscopy

Structured illumination microscopy (SIM) imaging approximately doubles the resolution both axially and laterally compared with conventional wide-field microscopy (Gustafsson 2000). For a typical fluorophore, it can resolve features separated by 110–130 nm in the lateral plane and 250 nm in the axial plane. A strong attribute is that 3D image acquisition is a routine and usually constitutive component of the SIM imaging process. SIM is also relatively easy to use, can be used on multiple standard fluorophores with no special sample preparation, and the newest platforms can do all this using live cells. For imaging of structures whose physical size lies in that range, SIM is possibly the most enticing super-resolution platform.

The principles underlying SIM are summarized very briefly in Fig. 3.2a. It involves the acquisition of several images of the same sample using light modulated by a periodic pattern of various orientations. The interference between the periodic pattern and the structure of the sample creates a Moiré pattern. This interference pattern contains downshifted high-frequency information. An algorithm is then used to reconstruct a single super-resolved image. For a far more detailed outline of the principles of structured illumination, the paper by Schermelleh (Schermelleh et al. 2008) is recommended.

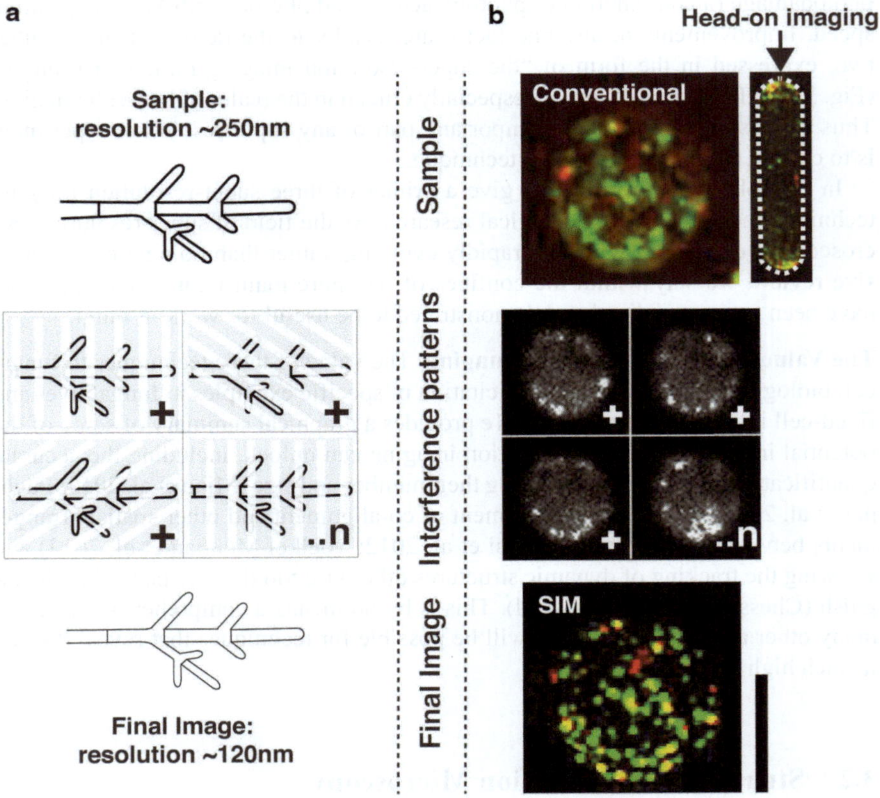

Fig. 3.2 Structured illumination microscopy (SIM) microscopy. **a** Principles underlying SIM microscopy. High spatial frequencies normally not assessable are collected by illuminating the sample with interference patterns created through a gridding pattern. Individual images are captured as the gridding device is shifted and rotated. The images are then processed by a computer algorithm to generate a super-resolved image. **b** Live-cell SIM versus conventional deconvolved wide-field images with an OMX BLAZE microscope (applied precision). Imaging Tea1-3mCherry (Tea1-3mCh, *red*) and Tea3-Green Fluorescent Protein (Tea3-GFP, *green*) polarity factors in the cell end of the fission yeast. Side-on cell was taken by conventional microscopy. Both head-on panels are maximum intensity projections of 2-μm-deep image z-stacks in the same cell (Chessel et al. Unpublished). The polarity factor clusters are resolvable to around 120 nm in size. Scale bar is 2 μm

A requirement of SIM is the acquisition of multiple "raw" images, which are then consequently processed to generate a single super-resolved image or stack. For this reason, acquiring a large number of "raw" images can limit SIM's applicability to live-cell microscopy mainly due to the effects of poor temporal resolution and potentially high photodamage. Newer commercial platforms are now available with very rapid image acquisition and are marketed for live-cell imaging. These still may not be suitable for very dynamic structures, whilst the very short exposure times required may not offer sufficient sensitivity for weak signals. SIM in most systems is a wide-field technique, and consequently acquiring good-quality images within thick specimens, where the object plane is a distance away from the coverslip, may be difficult (a limit of 10-μm distance from the coverslip is typical in one commercial system). Recently SIM has been combined with ultrathin planar illumination, which improves imaging within thick specimens and reduces the effects of high photodamage (Gao et al. 2012).

3.2.1 Applications in Mycological Imaging

Fixed Cell

SIM's relative ease of applicability extends to its compatibility with standard fixation and labelling methods found in conventional yeast microscopy. Structures that appear to be punctated by conventional microscopy can often be resolved into multiple different structures by super-resolution imaging. For example, 3D-SIM has been used to more accurately describe and quantitate densely arranged structures in yeast cells, such as linear actin cables stained with phalloidin in *Saccharomyces cerevisiae* (budding yeast) cells (Miao et al. 2013) and punctate polarity factors observed in *Schizosaccharomyces pombe* (fission yeast) cell ends labelled with a standard secondary antibody (Dodgson et al. 2013). Multicolour super-resolution microscopy can also allow a more exact determination of the nature of correlation between proteins or cellular structures. Even structures that appear to closely co-localize may no longer do so following a reduction in their PSFs by SIM. For example in budding-yeast vesicle assembly Gga2p-RFP (Red Fluorescent Protein) and β1-GFP, puncta overlap in conventional confocal imaging but are usually adjacent and distinct with 3D-SIM imaging (Daboussi et al. 2012). Also in budding yeast, more precise two-colour imaging has revealed at the organelle level that mitochondria are not associated with the endoplasmic reticulum (Swayne et al. 2011). Even between structures or proteins which still co-align with two-colour SIM, a reduction in their PSF size can now allow precise mapping of their respective spatial positions. For example, the protein SUMO was sufficiently resolved by 3D-SIM that it could now be positioned centrally within the budding-yeast synaptonemal complex (Voelkel-Meiman et al. 2013). Similarly, 3D-SIM was applied to precisely map the spatial layout of components of the replication factory with respect to their deoxyribonucleic acid (DNA) template (Saner et al. 2013).

Live Cell

Combining 2D-SIM with total internal reflection fluorescence (TIRF) microscopy can improve the compatibility of SIM with live-cell imaging (Dobbie et al. 2011). This has been used to image cortical proteins in live budding-yeast cells (Spira et al.

2012). As previously mentioned, the newest commercial systems have faster acquisition times and can realistically achieve live-cell 3D-SIM (lapses every ~1.2 s for a 2-μm-thick stack are possible in a commercial system). Commonly used tags such as GFP and mCherry are suitable, although they will be susceptible to photobleaching during the raw image acquisition. Figure 3.2b from our own work demonstrates live-cell 3D-SIM of labelled polarity factor clusters in a fission yeast cell end. The same polarity factor clusters can also be viewed by time-lapse imaging at 1-s intervals for around ten time points allowing the first direct tracking of these dynamic structures (Chessel Unpublished). Live-cell 3D-SIM has also been applied to map the proximity of mis-folded proteins to the budding-yeast vacuole membrane (Spokoini et al. 2012).

3.2.2 SIM Summary

SIM is arguably the least difficult super-resolution technique to apply, requiring little if any extra sample preparation compared to conventional microscopy and having, toward the same applicability in terms of 3D, multicolour and now live-cell imaging. Whilst the improvement in resolution is somewhat less than other super-resolution techniques, SIM is often advisable as a starting point before trying more difficult techniques. The achievable resolution might be sufficient for your imaging goals, and if the sample contains too much background to achieve good results, it will quickly become apparent. The extended raw image acquisition should not affect typical fixed-cell imaging (although anti-fade agents are advisable to reduce the effects of photobleaching) but may prove a limitation for live-cell imaging if the signal is weak, where increasing the illumination intensity leads to strong photobleaching and loss of signal. As a point of caution, it must be noted that structured illumination is susceptible to various artefacts resulting from factors including photobleaching, inaccurate system calibration, poor matching of the immersion oil refractive index to the sample leading to inaccurate or hard to interpret algorithmic reconstruction (Schaefer et al. 2004).

3.3 Stimulated Emission Depletion Microscopy

Unlike SIM or localization microscopy (LM), which are usually wide-field microscopy techniques, stimulated emission depletion (STED) microscopy is based on a confocal system (see Chap. 1). It might therefore be especially suited, with certain adaptations, to imaging of mycological samples growing in thick cultures where out of focus or scattered light would impair imaging quality (Gould et al. 2012). In yeast, live-cell STED with fluorescent proteins is now possible with resolution below 60 nm in the lateral plane (Stagge et al. 2013; Rankin et al. 2011).

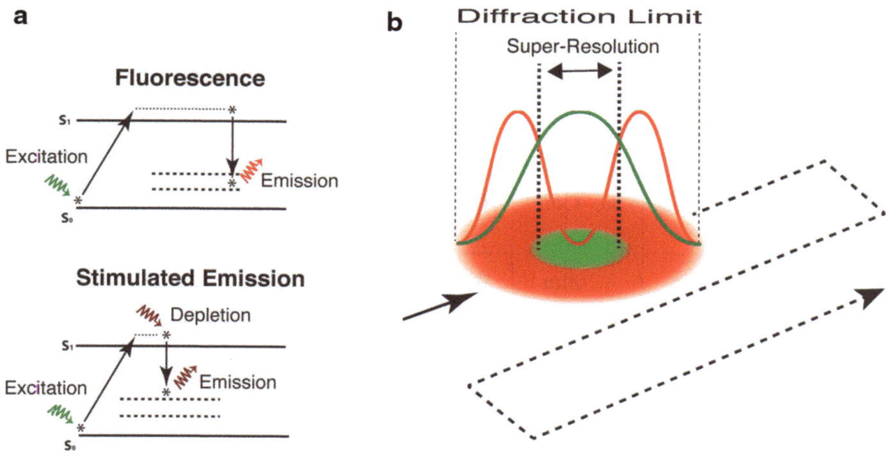

Fig. 3.3 Principles underlying stimulated emission depletion (STED) microscopy. **a** Highly simplified Jablonski diagram illustrating the transitions in electronic state during normal fluorescence and during stimulated emission. In normal fluorescence, a molecule is excited by light of shorter wavelength and then emits light of longer wavelength whilst transitioning back to an unexcited state. Stimulated emission of the excited molecular state by an additional depletion laser causes the emitted light to be of sufficiently longer wavelength and shorter fluorescent lifetime so that it can consequently be separated from normal fluorescence. **b** A doughnut-shaped depletion laser (in *red*) effectively reduces the point spread function (PSF) of normal fluorescence (in *green* and marked by a *double arrow*) below its diffraction limit since only the unstimulated light from the centre of the doughnut is collected. The sample is scanned with this arrangement as for conventional scanning confocal microscopy

STED is also different from SIM or LM in that generation of the super-resolved image is not finalized by a mathematical reconstruction but rather depends on a physical suppression of the effective PSF size (although deconvolution can be applied to improve image quality). This is summarized in Fig. 3.3 from an excellent review by Muller et al. 2012 which is also recommended for a much more in-depth description of the principles and methodology surrounding STED imaging. In brief, a diffraction-limited excitation spot is projected onto the sample together with a doughnut-shaped depletion beam that will restrict normal fluorescence to a central area smaller than the diffraction limit (Müller et al. 2012). Increasing the power of the depletion beam will increase the area of fluorophore depletion and further reduce the effective PSF. The depletion beam however also enhances photodamage and so is associated with signal loss that together can make STED imaging of weak signals difficult. It is also important to note that in conventional STED there is no impact on PSF in the axial plane, and consequently resolution is not improved along the *z*-axis (Farahani et al. 2010), although further implementations have been developed to partially rectify this (Laporte et al. 2013).

Table 3.1 Generalized relative attributes of the super-resolution techniques

System	Lateral Resolution	Axial Resolution	Platform	3D	Speed	Labels
SIM	~120nm	~250nm	Widefield	Routine	••	Standard fluorophores
STED	~60nm	No Gain	Confocal	Non-Routine	••• (Dependent on frame size)	Standard fluorophores
LM	~20nm	~20nm	Widefield	Non-Routine	•	Photoswitchable fluorophores

3.3.1 Applications in Mycological Imaging

Fixed Cell

STED has been mainly applied to immunofluorescently labelled fixed samples, and special fluorophores are usually required for STED. There are a range of commercially available dyes suitable for STED conjugated to secondary antibodies and with varied spectral properties (Table 3.1 of the review by Muller et al. 2012 recommends as a list of potential dyes). It should be noted that multicolour resolution most commonly requires two dyes which can be both depleted with the same laser.

Live Cell

Exploration of STED as a live-cell imaging technique has frequently used yeast as the imaging sample in proof-of-principle demonstrations. In the proof-of-principle application of STED to live cells, the vacuole membrane of *S. cerevisiae* cells was stained with the dye RH-414 (Klar et al. 2000). Staining with dyes as a method of live-cell labelling is very limited in application, but with fluorescently labelled proteins, there is much greater flexibility. For STED imaging in *S. cerevisiae,* this has been achieved by genetically fusing selected proteins with tags that bind to exogenous fluorescent labels (Stagge et al. 2013; Fitzpatrick et al. 2009) and by making use of new laser combinations to image GFP (Rankin et al. 2011). Using either method to image eisosomes at the cortex, lateral resolution below 60 nm was possible (Stagge et al. 2013; Rankin et al. 2011).

3.3.2 Stimulated Emission Depletion Microscopy Summary

Although yeast cells have been used as an exemplar sample in the development of STED, it is not yet widely used in mycology research for elucidating real biological questions. STED has potentially higher resolution than SIM in the lateral plane and will not generate artefacts from mathematical reconstructions. Fluorophore availability and high photodamage might be more limiting than other techniques, and as this is a confocal-based technique, the frame rate will scale with image size (or higher laser intensities must be used). Three-dimensional imaging is possible only with further techniques not yet commercially available (Wildanger et al. 2009). As suggested, it is possible that (with adjustment of the objective correction to correct for aberrations) STED will be most suitable for imaging of samples that exist in thick cultures (Gould et al. 2012).

3.4 Localization Microscopy

LM includes a group of related techniques such as photoactivated localization microscopy (PALM), fluorescence photoactivation localization microscopy (fPALM, see also Chap. 2), stochastic optical reconstruction microscopy (STORM) and ground state depletion microscopy followed by individual molecule return (GSDIM) based on the same principles outlined in Fig. 3.4a (Rust et al. 2006; Hess et al. 2006; Betzig et al. 2006, Fölling et al. 2008). By analysing the PSF of a single fluorophore, it is possible to determine its position with an accuracy dependent on the number of collected photons and the level of background. In standard experimental conditions, the PSF can then be substituted with a feature ten times smaller than the size of the PSF itself, allowing resolution well below the diffraction limit of the microscope.

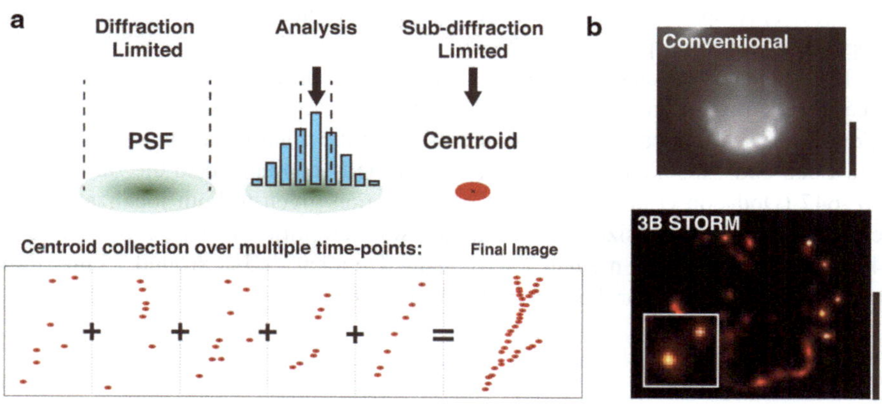

Fig. 3.4 Localization microscopy (LM). **a** Principles underlying LM. The point spread function *(PSF)* of a single fluorophore is diffraction limited in size but can be analysed to determine its centre to within an area below the diffraction limit. If emission is temporally controlled so that individual *PSFs* are distinguishable from each other, then individual points can be collected sequentially to generate a final super-resolved image. **b** Live-cell imaging of the polarity factor Tea3-Green Fluorescent Protein (Tea3-GFP) in the cell ends of fission yeast by conventional imaging and processed with 3B LM (our unpublished data). *Inset area* is double magnification showing individual clusters resolvable to 60 nm in size. Raw images taken on an OMX microscope. Scale bars 2 μm. *STORM* stochastic optical reconstruction microscopy

In biological samples, the PSFs of neighbouring fluorophores will usually overlap extensively rendering them inseparable. In LM, PSF overlap is avoided by temporal separation, with only a small fraction of all the fluorophores activated at any single time point. Over multiple time points (which may total several thousands in a typical LM experiment), a final super-resolved image is generated from the accumulated centroidal points. Activating only a subset of fluorophores in each time point is the key to LM, and the exact methods vary between different techniques. It is essential that both the PSF of the single fluorophore be detectable over the background (with enough photons to give a good localization accuracy) and that it be sufficiently distinct from other fluorophores excited at a given time point. A resolution of around 20 nm is possible in biological samples. Three-dimensional and multicolour imaging are possible, though not necessarily routine. Live-cell LM is difficult for similar reasons as live-cell SIM, in that the large number of individual image acquisitions required can lead to poor time resolution and to photodamage.

3.4.1 Applications in Mycological Imaging

Fixed Cell

In budding yeast, multicolour STORM imaging with photo-switchable dye pairs was used to image "barcoded" messenger ribonucleic acid (mRNA) transcripts (Lubeck and Cai 2012). In conventional microscopy, only low copy transcripts can be labelled because overlapping fluorophores cannot be differentiated. By contrast, STORM only sparsely activates overlapping fluorophores allowing much higher copy-number transcripts to be analysed.

LM with conventional fluorophores is now achievable with direct stochastic optical reconstruction microscopy (dSTORM; van de Linde et al. 2011). This technique was used to measure the size of clusters of the polarity factor Tea1 at fission yeast cell ends, labelled with a secondary antibody conjugated to the fluorophore AF-647 (Dodgson et al. 2013). With consideration given to both localization uncertainty and antibody size, the estimated resolution limit for those experiments of 42.5 nm was below the median Tea1 cluster size measured at around 75 nm, thus providing boundaries for the size of these structures.

Live Cell

Live-cell LM is possible with photo-switchable dyes (described in Chap. 2) or with photo-activatable proteins (Jones et al. 2011). In budding yeast, tagging of a PI3P reporter with mEos2 and live-cell PALM imaging allowed the maturation pattern of vesicles to be followed with multileveled detail (Puchner et al. 2013). With a resolution of 20 nm, vesicles with an average diameter of 80 nm could be accurately distinguished and measured in size. Furthermore, quantitative live-cell PALM was applied concurrently to molecule counting in the same vesicles. Molecules closely packed together could be individually counted, as during PALM imaging they are temporally separated. Such quantification required *in vivo* calibration methods that correct for over-counting from blinking and undercounting from non-maturation of the fluorophore. The number of molecules per vesicle was found to correlate with vesicle size as expected, and its measure contributed to understanding vesicle progression. A similar methodology was deployed to measure the number of CENP-A^{cnp1} tagged with mEos2 at the fission yeast centromere (Lando et al. 2012). Interestingly, relative deposition of CENP-A^{cnp1} molecules was found to peak in the later stages of the cell cycle. However, lack of agreement with other methods in the absolute counting of CENP-A^{cnp1} led to caution from the authors regarding the accuracy of their method for absolute measurements.

Three-dimensional LM is generally not routine, but by using a multifocal microscope in combination with PALM, Wisniewski et al. 2014 were able to image Cse4-tdEos at the budding-yeast centromere within a depth of ~4 μm and an axial resolution of ~50 nm. Using this approach, the Cse4 centromeric clusters were found to alter their size and shape during the cell cycle from a spherical grouping of ~400 nm in G1 to an ellipse grouping ~350 and ~200 nm in diameter.

Another live-cell LM technique is Bayesian analysis of blinking and bleaching (3B) LM (Cox et al. 2011), which is based on the natural blinking and bleaching properties of some fluorophores (such as GFP). This method creates a hidden Markov model of the dataset, including the blinking and bleaching properties of the fluorophores. This allows information to be extracted from data where many fluorophores are overlapping in each frame. The advantage of this approach is the relative ease by which it can be applied to imaging on standard microscopes with standard fluorophores. From our own 3B LM imaging, dynamic Tea3-GFP clusters at the fission yeast cell end could be tracked at a spatial resolution of 60 nm and a time resolution of 1.4 s (Fig. 3.4b).

In LM, because the final image is generated over multiple time points, there is a requirement for the target cell to be held very still over a prolonged period. Live fission yeast cells were immobilized with a microfluidic device that held and sustained the cells within a flow of media (Bell et al. 2014). Proof-of-principle PALM imaging of the centromere with mEos2-Cnp1 was then performed. The authors suggest that this method provides a more "natural" environment than trapping cells within gels for super-resolution imaging (Bell et al. 2014).

3.4.2 LM Summary

LM is the widest and fastest moving of the branches of super-resolution imaging. The sheer number of different techniques is vast and can seem bewildering. In biological samples, LM can achieve better resolution than SIM and STED, and 3D and multicolour imaging is becoming increasingly common. Live-cell PALM with protein tags like mEos2 is a particularly exciting technique, potentially requiring little sample preparation and extending to molecule counting as well as structure visualization.

3.5 Summary

The resolution limit of the conventional microscope has been a major barrier to the ability of biologists to accurately describe the detailed mechanisms of cells. The advent of super-resolution microscopy has and is increasingly pushing this boundary back in what represents an imaging revolution. Previously indistinguishable structures can be identified, accurately measured and tracked and the spatial relationships between protein pairs more closely described, and now even the quantification of proteins within a structure is possible.

Beginning with the biological question of interest that might be answered by super-resolution imaging, choose a technique that will most directly provide the solution. This review has covered the relative attributes of SIM, STED and LM imaging (generalized in Table 3.1), but in reality the relative performance and applicability of a technique are going to be highly experiment specific and may involve considerations and compromises not featured in this review. SIM is the most straightforward technique to apply, with 3D, multicolour and time-lapse imaging as a routine feature. SIM is not usually suitable for imaging in thick specimens and has less resolution than STED and LM. STED is a confocal-based technique so could be more suitable for imaging far away from the coverslip, and its potential lateral resolution is better than SIM. 3D and multicolour STED imaging are not yet routine, and the frame rate in time-lapse imaging depends on the image size. LM imaging has potentially the best resolution of all the techniques, but 3D and multicolour imaging within thick specimens may be difficult. Be aware that super-resolution fixed- and live-cell imaging represent two very different sized challenges with all three techniques associated with a potential decrease in speed and photostability that might render it difficult with live-cell imaging.

The uptake of super-resolution in the mycological field has arguably been slow, with a similar paper in the bacterial field able to review from a far larger group of experimental papers (Coltharp and Xiao 2012). Any lack of uptake is presumably the result of unfamiliarity with new techniques and maybe more because super-resolution platforms, although commercially available, are still fairly scarce, sometimes expensive and certainly less available than conventional ones. Super-resolution microscopy is a new technology in the early adoption phase, so if as likely you are not in an institution with the suitable instruments for your aims, then be prepared to be proactive in searching for potential hosts and collaborations. It is advisable to establish contacts and gain advice from imaging experts within your own institution and any visiting speakers with experience of super-resolution imaging. Be persistent in approaching the relevant people or groups with access to the relevant systems, and once an understanding is reached, then you will probably have to travel to the site of the super-resolution platform. Your hosts or collaborators are also likely to be experts in this field and a valuable source of advice.

For decades the diffraction limit has stood before the biological investigator as a barrier to be aware of but not to be overcome. The dawn of the super-resolution microscope has now broken this limit and represents an opportunity to describe your biological system with new radical detail.

Acknowledgments We would like to thank Nicola Lawrence and Alex Sossick for assistance with imaging and Eric Betzig for use of the super-resolution imaging-limitation triangle. This work was supported by an European Research Council (ERC) Starting Researcher Investigator Grant (R.E.C.-S., J.D.; SYSGRO), a Human Frontier Science Program (HFSP) Young Investigator Grant (R.E.C.-S., A.C., J.D.; HFSP RGY0066/2009-C), Biotechnology and Biological Sciences Research Council (BBSRC) Responsive Mode grant (R.E.C.-S., A.C., J.D.; BB/K006320/1) and an MRC grant (S.C.). S.C. was also supported by a Royal Society University Research Fellowship.

References

Bell L, Seshia A, Lando D, Laue E, Palayret M, Lee SF, Klenerman D (2014) A microfluidic device for the hydrodynamic immobilisation of living fission yeast cells for super-resolution imaging. Sens Actuators B Chem 192:36–41

Betzig E, Patterson GH, Sougrat R, Lindwasser OW, Olenych S, Bonifacino JS et al (2006) Imaging intracellular fluorescent proteins at nanometer resolution. Science 313(5793):1642–1645

Chessel A, Boussier J, Dodgson J, Carazo-Salas RE (Unpublished) Investigating single particle cortical polarity nanocluster dynamics without tracking

Coltharp C, Xiao J (2012) Superresolution microscopy for microbiology. Cell Microbiol 14(12):1808–1818

Cox S, Rosten E, Monypenny J, Jovanovic-Talisman T, Burnette DT, Lippincott-Schwartz J et al (2011) Bayesian localization microscopy reveals nanoscale podosome dynamics. Nat Methods 9(2):195-200. doi:10.1038/nmeth.1812

Daboussi L, Costaguta G, Payne GS (2012) Phosphoinositide-mediated clathrin adaptor progression at the trans-Golgi network. Nat Cell Biol 14(3):239–248

Dobbie IM, King E, Parton RM, Carlton PM, Sedat JW, Swedlow JR et al (2011) OMX: a new platform for multimodal, multichannel wide-field imaging. Cold Spring Harb Protoc 2011(8):899–909

Dodgson J, Chessel A, Yamamoto M, Vaggi F, Cox S, Rosten E et al (2013) Spatial segregation of polarity factors into distinct cortical clusters is required for cell polarity control. Nat Commun 4:1834–1839 (Nature Publishing Group. 1AD)

Farahani J, Schibler M, Bentolila L (2010) Stimulated emission depletion (STED) microscopy: from theory to practice. Microscopy: Science, Technology, Applications and Education

Fölling J, Bossi M, Bock H, Medda R, Wurm CA, Hein B et al (2008) Fluorescence nanoscopy by ground-state depletion and single-molecule return. Nat Methods 5(11):943–945

Fitzpatrick JAJ, Yan Q, Sieber JJ, Dyba M, Schwarz U, Szent-Gyorgyi C et al (2009) STED nanoscopy in living cells using Fluorogen Activating Proteins. Bioconjug Chem 20(10):1843–1847

Gao L, Shao L, Higgins CD, Poulton JS, Peifer M, Davidson MW et al (2012) Noninvasive imaging beyond the diffraction limit of 3D dynamics in thickly fluorescent specimens. Cell 151(6):1370–1385

Gould TJ, Burke D, Bewersdorf J, Booth MJ (2012) Adaptive optics enables 3D STED microscopy in aberrating specimens. Opt Express 20(19):20998–21009

Gustafsson MG (2000) Surpassing the lateral resolution limit by a factor of two using structured illumination microscopy. J Microsc 198(Pt 2):82–87

Hess ST, Girirajan TPK, Mason MD (2006) Ultra-high resolution imaging by fluorescence photoactivation localization microscopy. Biophys J 91(11):4258–4272

Jones SA, Shim S-H, He J, Zhuang X (2011) Fast, three-dimensional super-resolution imaging of live cells. Nat Methods 8(6):499–508

Klar TA, Jakobs S, Dyba M, Egner A, Hell SW (2000) Fluorescence microscopy with diffraction resolution barrier broken by stimulated emission. Proc Natl Acad Sci U S A 97(15):8206–8210

Lando D, Endesfelder U, Berger H, Subramanian L, Dunne PD, McColl J et al (2012) Quantitative single-molecule microscopy reveals that CENP-A(Cnp1) deposition occurs during G2 in fission yeast. Open Biol 2(7):120078

Laporte GPJ, Conkey DB, Vasdekis A, Piestun R, Psaltis D (2013) Double-helix enhanced axial localization in STED nanoscopy. Opt Express 21(25):30984

Lubeck E, Cai L (2012) Single-cell systems biology by super-resolution imaging and combinatorial labeling. Nat Methods 9(7):743–748

Miao Y, Wong CCL, Mennella V, Michelot A, Agard DA, Holt LJ et al (2013) Cell-cycle regulation of formin-mediated actin cable assembly. Proc Natl Acad Sci U S A 110(47):E4446–E4455

Müller T, Schumann C, Kraegeloh A (2012) STED microscopy and its applications: new insights into cellular processes on the nanoscale. Chemphyschem 13(8):1986–2000

Puchner EM, Walter JM, Kasper R, Huang B, Lim WA (2013) Counting molecules in single organelles with superresolution microscopy allows tracking of the endosome maturation trajectory. Proc Natl Acad Sci U S A 110(40):16015–16020

Rankin BR, Moneron G, Wurm CA, Nelson JC, Walter A, Schwarzer D et al (2011 Jun 22) Nanoscopy in a living multicellular organism expressing GFP. Biophys J 100(12):L63–L65

Rust MJ, Bates M, Zhuang X (2006) Sub-diffraction-limit imaging by stochastic optical reconstruction microscopy (STORM). Nat Methods 3(10):793–795

Saner N, Karschau J, Natsume T, Gierlinski M, Retkute R, Hawkins M, et al (2013) Stochastic association of neighboring replicons creates replication factories in budding yeast. J Cell Biol 202(7):1001–1012

Schaefer LH, Schuster D, Schaffer J (2004) Structured illumination microscopy: artefact analysis and reduction utilizing a parameter optimization approach. J Microsc 216(Pt 2):165–174

Schermelleh L, Carlton PM, Haase S, Shao L, Winoto L, Kner P et al (2008 Jun 6) Subdiffraction multicolor imaging of the nuclear periphery with 3D structured illumination microscopy. Science 320(5881):1332–1336

Spira F, Mueller NS, Beck G, Olshausen von P, Beig J, Wedlich-Soldner R (2012) Patchwork organization of the yeast plasma membrane into numerous coexisting domains. Nat Cell Biol 14(6):640–648. doi:10.1038/ncb2487

Spokoini R, Moldavski O, Nahmias Y, England JL, Schuldiner M, Kaganovich D (2012) Confinement to organelle-associated inclusion structures mediates asymmetric inheritance of aggregated protein in budding yeast. Cell Rep 2(4):738–747

Stagge F, Mitronova GY, Belov VN, Wurm CA, Jakobs S (2013) Snap-, CLIP- and Halo-Tag labelling of budding yeast cells. PLoS One 8(10):e78745

Swayne TC, Zhou C, Boldogh IR, Charalel JK, McFaline-Figueroa JR, Thoms S et al (2011) Role for cER and Mmr1p in anchorage of mitochondria at sites of polarized surface growth in budding yeast. Curr Biol 21(23):1994–1999

van de Linde S, Löschberger A, Klein T, Heidbreder M, Wolter S, Heilemann M et al (2011) Direct stochastic optical reconstruction microscopy with standard fluorescent probes. Nat Protoc 6(7):991–1009

Voelkel-Meiman K, Taylor L, Mukherjee P (2013) SUMO localizes to the central element of synaptonemal complex and is required for the full synapsis of meiotic chromosomes in budding yeast. PLoS Genet 9(10):e1003837. doi:10.1371/journal.pgen.1003837

Wildanger D, Medda R, Kastrup L, Hell SW (2009) A compact STED microscope providing 3D nanoscale resolution. J Microsc 236(1):35–43

Wisniewski J, Hajj B, Chen J, Mizuguchi G, Xiao H (2014) Imaging the fate of histone Cse4 reveals de novo replacement in S phase and subsequent stable residence at centromeres. ELife (Europe PMC Article—Europe PubMed Central)

Chapter 4
Fourier Transform Infrared (FTIR) Microscopy and Imaging of Fungi

Annette Naumann

4.1 Introduction

Advances in understanding fungal development and interactions with inorganic and organic substrates or living organisms require spatially localised information on molecular composition in relation to physical structures revealed in microscopic images. Sample pretreatment for chemical analyses, such as gas chromatography mass spectrometry (GC-MS) or liquid chromatography mass spectrometry (LC-MS), frequently involves sample homogenisation to guarantee that the small measured subsample represents the larger sample. Such treatment disintegrates the anatomical structure and cell compartmentalisation and the associated chemical reactions can change the molecular composition. Regardless of analytical technique, reducing the size of subsamples to achieve a higher spatial resolution requires a larger number of measurements. Locating subsamples within the sample and correlating them with the anatomical structure can become a challenge with increasing subsample number. The sensitivity of the analytical method limits the minimal possible size of subsamples. Alternatively, advanced light or electron microscopy enables higher spatial resolution, but imaging the distribution of single or a few selected chemicals within the sample usually requires dyes or labels. Such treatment can cause redistribution of soluble components or chemical reactions that perturb the original sample composition and may be limited by specificity and accessibility of binding sites. In

A. Naumann (✉)
Federal Research Centre for Cultivated Plants, Julius Kühn-Institute,
Königin-Luise-Str. 19, 14195 Berlin, Germany

Molecular Wood Biotechnology and Technical Mycology,
Forest Botany and Treephysiology, Georg-August-University Göttingen,
Büsgen-Institute, Büsgenweg 2, 37077 Göttingen, Germany
e-mail: anauman1@gwdg.de

T. E. S. Dahms and K. J. Czymmek (eds.), *Advanced Microscopy in Mycology*,
Fungal Biology, DOI 10.1007/978-3-319-22437-4_4

contrast, Fourier transform infrared (FTIR) spectroscopy in combination with microscopy generates spatially resolved spectral information and enables imaging the distribution of a multitude of chemical constituents. In this way, chemical imaging by FTIR microscopy provides direct insight into the sample composition without perturbing by staining or labelling (Salzer et al. 2000).

4.2 Basic Principles of FTIR Microscopy and Imaging

In the following section, FTIR spectroscopy, microscopy and imaging are briefly outlined based on Naumann et al. (2007) and Fackler and Thygesen (2013), dealing with FTIR microscopy of wood and its degradation by fungi, as well as on detailed information in Salzer et al. (2000), Chalmers and Griffiths (2002), Günzler and Gremlich (2002) and Salzer and Siesler (2009, 2014).

4.2.1 FTIR Spectroscopy

For FTIR analysis, infrared (IR) light passes through an interferometer, irradiates a sample, and causes vibrations of polarised molecular bonds, such as O–H, N–H, C–H, C = O. Specific molecular vibrations absorb IR light at distinct wavenumbers (wavenumber [cm^{-1}] = 10^7/wavelength [nm]) of the FTIR spectrum, allowing absorption bands to be assigned to their chemical basis. In simple molecules, each absorption band can be assigned to a distinct vibrational mode of a specific bond. Stretch vibrations change bond lengths and deformation vibrations change bond angles. In complex materials, such as fungi and all other biological systems, absorption bands of many molecules overlap and complicate spectral interpretation. However, the highly complex FTIR spectra of biomaterials provide an informative qualitative overview of the main components such as lipids, proteins, nucleic acids and carbohydrates (Fig. 4.7). Comparison with reference spectra of pure substances, such as ergosterol, glucose, glucans and chitin, enables tentative assignment of FTIR bands to individual chemical substances. For this purpose, spectra and band assignments are compiled in original articles (Mohaček-Grošev et al. 2001), reviews (e.g. Naumann 2000, Schulz and Baranska 2007, Movasaghi et al. 2008), printed spectra atlases (e.g. Pachler et al. 1988), and digital spectral libraries (e.g. NIST Chemistry WebBook 2014).

Moreover, absorbance of FTIR spectra (band height and area) is related to the concentration of the corresponding chemical substances, which enables semiquantitative analyses of relative concentrations. Absolute quantification of chemical components within biomaterials can be achieved by measuring the target component with varying concentrations in a set of reference samples using a reference method (e.g. HPLC, GC-MS, LC-MS etc.). By applying a partial least square regression (PLSR) algorithm, the resulting reference values can be used to establish a calibration model for predicting the concentration of unknown samples. Numerous examples for quantification of valuable substances in plant material are given in Schulz and Baranska (2007). Sensitivity varies for specific substances and biomaterials, but generally the detection limit is in the mg/g range (Schulz and Baranska 2007). Mass spectrometric analyses are potentially more sensitive, but FTIR analyses provide advantages regarding sample preparation, range of spatial resolution, analysis speed and costs.

4.2.2 Classification Methods

The complexity of FTIR spectra represents the immense diversity of molecules in biomaterials as a highly specific fingerprint of their chemical composition. As such, pattern recognition methods are necessary to classify FTIR spectra according to their spectral and the corresponding chemical similarity or heterogeneity (Salzer et al. 2000; Mariey et al. 2001; Krafft et al. 2009; Salzer and Siesler 2014). In this way, the diversity within or between samples (individuals of close or distant taxonomic relationship) can be studied, as well as the influences of abiotic or biotic factors.

Statistical methods for pattern recognition can be divided into two main groups called unsupervised and supervised methods (Salzer et al. 2000; Mariey et al. 2001; Krafft et al. 2009; Salzer and Siesler 2014). Samples for which only the FTIR spectra are available require unsupervised methods that group the spectra solely based on the FTIR data without a priori knowledge. Examples are principle component analysis (PCA) and cluster analysis (e.g. hierarchical, k-means, C-means). PCA reduces the complexity of the spectral data without decreasing their variance and can be combined with other pattern recognition methods. Unsupervised methods can produce novel, unforeseen results (Salzer et al. 2000). The wavenumbers of the spectral differences revealed by unsupervised classification methods can give valuable hints on their chemical nature and on choosing adequate analytical methods for verifying their exact chemical structure. In this way, unsupervised methods identify spectra of high chemical similarity and heterogeneity and are a valuable tool to assign spectra to classes of typical chemical composition (e.g. fungal taxa) or sample areas containing specific components. Predefined classes are essential to refine the results by applying supervised methods. Alternatively, class assignment can use knowledge of the chemical composition obtained through chemical analyses or ideally combine all available a priori knowledge about the sample.

Supervised methods assign spectra to such predefined classes. Based on spectra with known class membership (training data), a prediction model is built to assign unknown spectra to these classes. Examples are discriminant analysis, soft independent modelling of class analogy (SIMCA), artificial neural network analysis and support vector machines (Salzer et al. 2000; Mariey et al. 2001; Krafft et al. 2009; Salzer and Siesler 2014). Validation of these methods is possible by using an independent test set of spectra or the “leave one out” method, where each sample is removed in turn and the remaining used to evaluate the classification quality (Mariey et al. 2001).

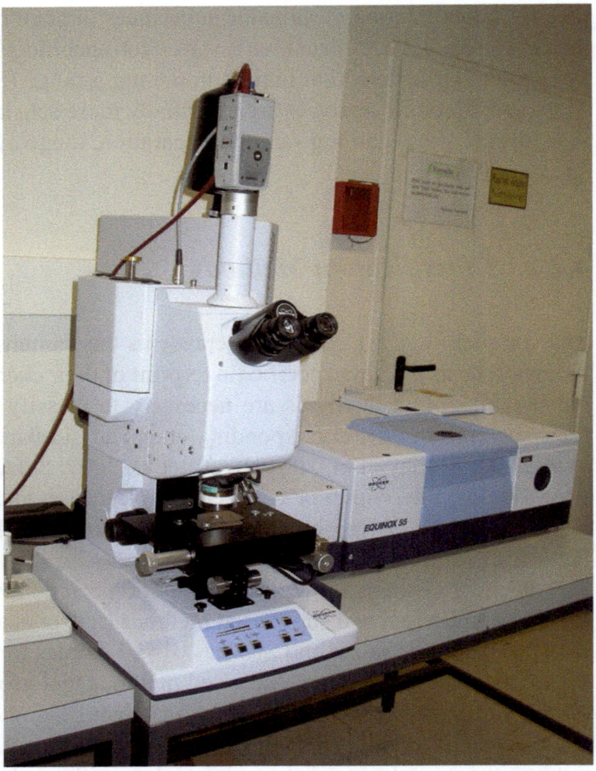

Fig. 4.1 Fourier transform infrared (FTIR) microscope Hyperion 3000 (Bruker Optics, Ettlingen, Germany) equipped with a single-channel detector and a 64 × 64 focal plane array (FPA) detector

4.2.3 FTIR Microscopy

In FTIR microscopy (Figs. 4.1 and 4.2), also referred to as microspectroscopy or spectromicroscopy, radiation of an IR source (e.g. globar) transmitted through a sample or reflected at the surface irradiates a detector (Salzer and Siesler 2009, 2014). Switching between IR and visible radiation enables control of the measurement area and taking visible light microscopic images for comparison with IR measurements. The detected IR range depends on the type of detector and manufacturer.

4.2.4 Sample Preparation

With an FTIR microscope (Figs. 4.1 and 4.2), samples are frequently measured in transmission mode on IR-transmissive windows such as potassium bromide (KBr) for dry samples, zinc selenide or calcium fluoride for wet samples. Sample thickness must be adjusted to allow sufficient transmittance and absorbance, usually between 5 and 15 µm. Tissue of fruiting bodies or thick mycelium requires sectioning

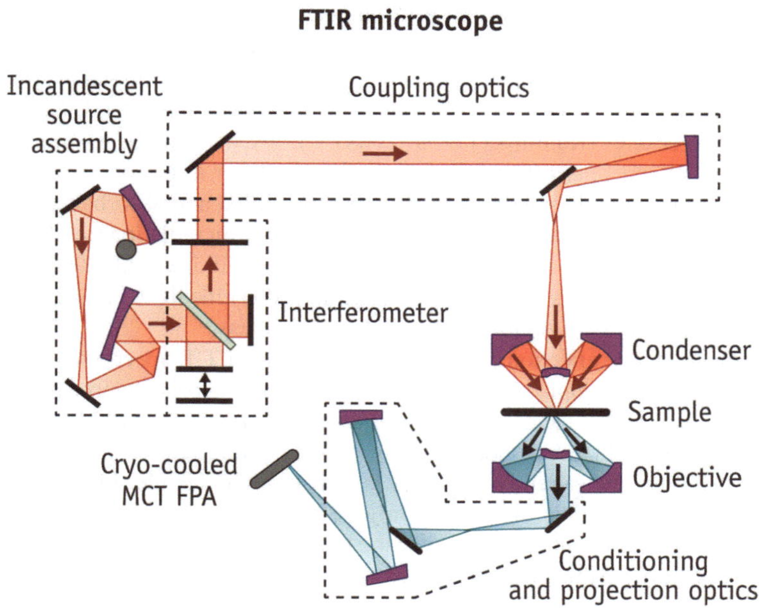

Fig. 4.2 Schematic representation of an FTIR microscope (Rowlette et al. 2014). *FTIR* Fourier transform infrared, *MCT FPA* mercury cadmium telluride focal plane array. (Reprinted with permission from the March/April 2014 edition of *BioOptics World* Copyright 2014 by PennWell)

(Pallua et al. 2012; Posch et al. 2013). In the case of nonadhesive, curling sections, a weight on top during drying helps achieve a flat sample surface. Care must be taken to change the chemical sample composition as little as possible and therefore cryo-sectioning is preferable. For cryo-sectioning, the sample is fixed in a freezing drop of water before cutting sections. Liquids containing glycols and resins can fix samples more firmly, but can also contaminate the sections. Note that there can be redistribution of sample liquid and solid fragments due to sectioning and freezing. Fixing (e.g. formalin) and embedding (e.g. paraffin) chemicals increase the risk of chemical reactions and redistribution, altering the original chemical composition. In difficult samples, embedding might be the only way to obtain sections with intact morphology and valuable information can be gained from such samples if their chemistry is stable. However, special care has to be taken to rule out investigation of artefacts by monitoring chemical changes during embedding. Sample preparation of fresh, dried, cryopreserved, chemically fixed and embedded tissues and cells for IR spectroscopy and imaging is described and compared in detail in Lyng et al. (2011).

Single hyphae are close to the spatial detection limit of IR microscopy. Whether they can be measured depends on their diameter and density-related absorbance and the instrument type used (see below). Mycelia with sparse spatial distribution of single hyphae can be measured directly without sectioning.

Simple reflection at the surface of usually nonreflecting biomaterials is often insufficient to achieve satisfactory spectral quality. To obtain sufficient absorbance of thin single hyphae, reflection mode using a gold mirror or IR-reflective slides (e.g. MirrIR low-e microscope slides, Kevley Technologies, Chesterland, Ohio, USA) has been successful (Kaminskyj et al. 2008). In this way, the beam transmits the sample before and after reflection at the reflective layer in a process called transflectance. Thus, sample thickness is effectively doubled and absorbance increased accordingly. However, near hyphal tips, even synchrotron radiation (see below) and transflectance produced spectra with low absorbance (Jilkine et al. 2008). To grow hyphae across IR-reflective slides, Jilkine et al. (2008) incubated spore inoculated agar and subsequently lyophilised the hyphae to dryness.

Attenuated total reflection (ATR) objectives enable FTIR spectra to be recorded from nonreflecting sample surfaces in many cases without prior sample preparation, referred to as micro-ATR analysis (Kazarian and Chan 2010, see below). For thin hyphae, visible light microscopy usually produces a structured image necessary for localising the measurement point or area within the sample. In thick samples, the corresponding visible image can only be taken in reflection mode and might not reveal sufficient contrast to localise the measurement area precisely. Most ATR objectives do not permit simultaneous IR and visible microscopy. Pressing the ATR crystal on the sample may cause a slight movement of the sample so that the sample areas in the visible and IR image no longer exactly correspond. Localising the measurement area can also be a challenge in thin, but optically homogeneous samples. If the ATR crystal contacts various measurement areas consecutively, there can be cross-contamination between the measured areas.

Generally, the strong IR absorbance of water obscures many minor bands of secondary metabolites. Therefore, working with dry samples is always preferable if drying does not impair the results. However, aqueous or living samples (enzymatic studies) can produce valuable results if humidity is strictly controlled (Gierlinger et al. 2008). For main bands, direct spectral interpretation is an option, but minor bands require subtraction of a background for humidity correction.

4.2.5 FTIR Mapping

In FTIR microscopy, there are two ways to generate images, each using different detector types, called FTIR mapping and FTIR imaging, respectively (Salzer and Siesler 2014). Single-element mercury cadmium telluride (MCT) detectors record one spectrum at a time, usually covering a spectral range from 300 or 600 to 4000 cm^{-1}. Apertures determine the size of the measurement area, usually between 10×10 and 100×100 μm. The signal-to-noise ratio declines with decreasing aperture, hence the need for synchrotron light. Moving the sample in small steps (e.g. 5–20 μm) and recording spectra of relatively small sizes (e.g. 10×10–50×50 μm), with a single-channel detector after each step, produces so-called mappings (Jilkine et al. 2008). Height or area of each selected band in the spectrum can be used to visualise the semiquantitative distribution of the assigned chemical component in grey or pseudo-colour scale by means of suitable computer software (e.g. OPUS software, Bruker Optics, Ettlingen, Germany; Jilkine et al. 2008). Such mappings usually require several hours to days and are progressively replaced by the faster FTIR-imaging technique. Nevertheless, single-element measurements are a valuable tool for getting a fast overview of the main composition of different samples or sample areas (Naumann et al. 2005).

4.2.6 FTIR Imaging

In contrast to consecutive mapping, FTIR imaging, with modern multielement detectors consisting of numerous single-detector elements, records thousands of spectra simultaneously within a few minutes (Salzer and Siesler 2014). For example, 16×16, 32×32, 64×64, or 128×128 single-MCT detector elements are combined in focal plane array (FPA) detectors typically covering spectral ranges about 900–4000 cm^{-1}. The size of single-FPA detector elements is producer-specific, for example 4×4 or 6.25×6.25 μm, and is close to the wavenumber-dependent spatial resolution limit. At 1500 cm^{-1} and higher wavenumbers, a theoretical spatial resolution of 4 μm and better is achieved. Decreasing wavenumbers impair spatial resolution, theoretically to 7 μm at 900 cm^{-1}. The size of measurement area ranges from approximately 90×90 to 700×700 μm and is related to detector size and producer-dependent microscope configuration. Linear arrays of, for example, 16 (2×8) single-detector elements combine features of mapping and imaging: 16 spectra are recorded simultaneously, which is repeated consecutively with samples on a stepwise moving stage.

4.2.7 Synchrotron FTIR Microscopy

Synchrotron light is considerably brighter than light from a globar (Kaminskyj et al. 2008) and can be used for mapping with a single-element detector as well as for imaging with an FPA detector. Spectral quality improves with increasing brightness. Especially in the case of single, thin hyphae (Kaminskyj et al. 2008), synchrotron FTIR microscopy might make the difference in spectral quality that ultimately helps solve an experimental question (see Chap. 8). However, with the increasing sensitivity of modern FTIR microscopes, systems with a thermal globar source and FPA detector produce valuable information on the chemical composition of many types of specimens, even single hyphae (Isenor et al. 2010). The increasing numbers and areas of applications of globar/FPA systems open the way for FTIR imaging to become a mainstream laboratory-based method. In contrast, access to synchrotron sources is restricted and synchrotron FTIR microscopy requires specialised expertise.

4.2.8 Micro- and Macro-ATR Analysis

Micro-ATR objectives record FTIR spectra in direct contact between the sample surface and the ATR crystal (e.g. germanium; Kazarian and Chan 2010). An advantage of such measurements over FTIR in transmission mode is the ability to measure the composition of the surface of thick samples without laborious sectioning. For spectral detection, single-element, FPA and linear-array detectors are suitable. Germanium crystals in micro-ATR objectives potentially enhance spatial resolution to ~2 μm, analogous to an oil immersion objective, but the magnification reduces the field of view (Kazarian and Chan 2010).

For ATR measurements (Günzler and Gremlich 2002; Chalmers and Griffiths 2002; Kazarian and Chan 2010), the ATR crystal (also referred to as internal reflection element) and the sample are pressed tightly together. At this interface, the IR beam is totally reflected and attenuated by absorbance in the sample surface based on its chemical composition (Figs. 4.3 and 4.4).

Complementary to all FTIR-microscopic techniques, macro-ATR spectroscopy (Fig. 4.3) can serve as a valuable tool for fast measurement (usually < 1 min) of spectra from reference or bulk material, including liquids and solids. The typical

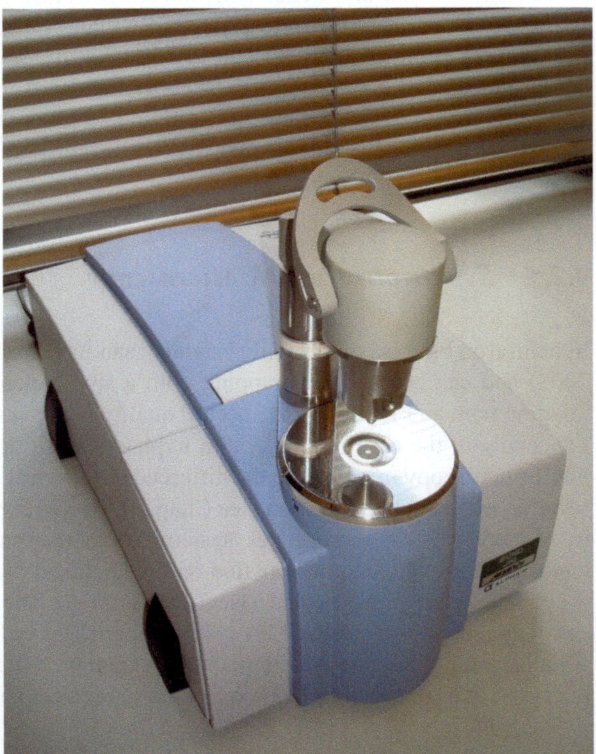

Fig. 4.3 Fourier transform infrared attenuated total reflection (FTIR-ATR) spectrometer Alpha (Bruker Optics, Ettlingen, Germany)

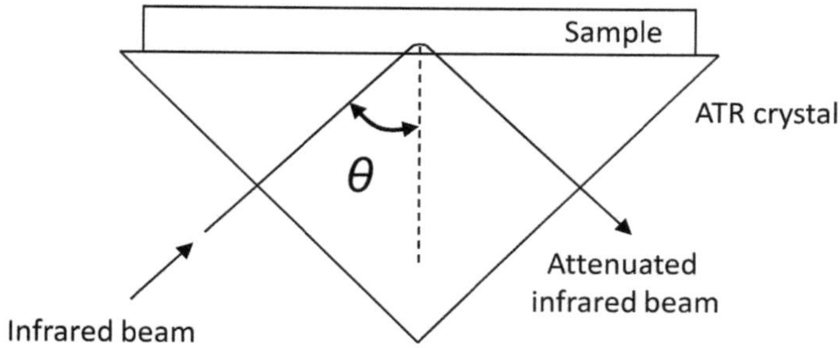

Fig. 4.4 Schematic representation of attenuated total reflection. *θ* angle of incidence, *ATR* attenuated total reflection. (Modified from Günzler and Gremlich 2002)

ATR crystal (d=~1 mm) can be used to record spectra at various sites of the sample surface. Further, ground samples can easily be compared to identify significant differences on the whole tissue level, which are laborious to detect on the microscopic scale. For larger samples in the centimetre range, macro-ATR imaging is a promising technique, able to achieve configuration-dependent spatial resolutions of approximately 50–500 µm. For this purpose, a larger ATR crystal serves as a sample stage and is combined with a linear or FPA detector (Kazarian and Chan 2010). Thus, macro-ATR imaging can fill the gap between single-point macro-ATR measurements and FTIR imaging, increasing the field of view. In the same way, macro-imaging in transmission or reflection modes is possible (Salzer and Siesler 2009).

4.2.9 Spectral Preprocessing

Spectra analyses usually require spectral preprocessing to account for impacts such as baseline shifts, carbon dioxide, humidity, light scattering and absorbance deviation due to factors such as varying sample thickness or contact between the ATR crystal and sample. Frequent preprocessing methods are baseline, carbon dioxide and humidity corrections, (vector) normalisation, extended multiplicative signal correction (EMSC) and first or second order derivatisation (Salzer and Siesler 2009, 2014).

4.3 Chemical Imaging of Fungi

For meaningful chemical imaging, detailed assignment of FTIR bands to specific vibrations of molecular bonds and chemical compounds is necessary. A comprehensive overview of FTIR bands from fungal fruiting bodies and spores is given in Table 4.1 and chitin, α- and β-glucan bands in more than 70 species are compared in Mohaček-Grošev et al. (2001). Extensive band assignments for marine yeasts are compiled in Vongsvivut et al. (2013) and compositional information on extracted fungal cell walls is outlined in Michell and Higgins (2002).

Table 4.1 Infrared spectral bands (4000–800 cm^{-1}) observed for a fruiting body and spores from the button mushroom *Agaricus bisporus*. (Reproduced with permission from Mohaček-Grošev et al. 2001 © Elsevier)

Spores	Cap skin	Cap	Gills	Stalk	Assignment
3330 vs, vbr	3284 vs, vbr	3280 vs, vbr	3356 vs, vbr	3412 vs, vbr	O–H stretching
				3288 vs, sh	O–H stretching
3012 m			3008 m		Aromatic C–H stretching (pigment)
	2972 ms, sh				
2954 ms	2956 s	2956 s	2956 s	2958 s	C–H stretching (pyranose ring)
	2938 s	2938 s			C–H stretching (chitin)
2928 s	2922 s	2924 s	2926 s	2924 s	ν_{as} (CH_2) lipid + C–H stretching (pyranose ring)
2874 m	2882 ms, sh	2882 m	2872 m	2878 m	CH_2 stretching (pyranose ring)
2856 ms	2854 ms	2856 m	2856 m	2852 m	ν_s(CH_2) lipid + C–H stretching (pyranose ring)
1747 m	1736 m, sh		1744 m	1742 w, sh	C=O lipid ester
1642 vs	1654 vs, br	1656 vs, vbr	1650 vs, br	1650 s	Protein Amide I
			1634 vs	1634 s	+ O–H–O bending of bound water
1564 s	1560 s, br	1566 s, br	1570 s, sh	1566 s, br	Chitin amide band
1514 m, sh			1518 m, sh		Protein Amide II
1462 m	1460 s	1458 ms	1458 m		CH_2 bending (pyranose ring)
	1434 s			1448 m, br	
1416 m		1410 ms, br		1410 m, br	C–H bending +
1400 m			1402 m		C–O–H bending (pyranose ring)
1378 m	1378 s, br	1380 ms	1380 m	1378 m	
	1352 m	1352 m, sh			
1310 m, br	1320 m	1322 m		1316 m	
			1296 m, br		C–O–H bending + CH_2 deformation (pyranose ring)
1260 m	1264 m	1264 m		1264 mw	
			1252 m, sh		
1242 m	1234 mw, sh	1240 m, sh		1236 mw	
1204 mw	1200 mw	1202 w	1204 mw	1206 mw	C–C stretching + C–O stretching + C–H deformation (pyranose ring)
1155 m	1150 mw	1154 m	1154 m	1154 m	
	1112 s, sh	1110 ms, sh	1106 m, sh	1112 m	
1078 vs, br	1084 vs	1084 vs	1076 ms	1074 s, br	
	1050 s	1054 s			
1032 s	1040 s		1036 m	1034 s	C–C stretching
	1022 s	1022 s			
	992 m, sh				
	992 m, sh	972 w, sh		974 w, sh	
952 w	954 m	952 mw	954 w	950 w	Glucan band (β-anomer C–H deformation)
930 w, sh	928 m	930 mw			Glucan band (α anomer C–H deformation)
888 w, br	886 m	884 w	888 w	900 w	Glucan band (β anomer C–H deformation)
	874 w, sh	876 w, sh			Mannan band
834 w, sh					Glucan band (α-anomer C–H deformation)
806					Mannan band

Abbreviations of relative band intensity and shape: *s* strong, *m* medium, *w* weak, *v* very, *br* broad, *sh* sharp

4.3.1 Single Hyphae

To demonstrate spatial differences in cell composition of hyphae, recording single-synchrotron FTIR spectra at different distances from hyphal tips provided a powerful tool (Szeghalmi et al. 2007). Two Ascomycetes (*Aspergillus nidulans* and *Neurospora*) and a Zygomycete (*Rhizopus*) were compared showing that the cytoplasm near the apical tip contained more protein- and lipid-membrane-rich organelles. Carbohydrate-band intensity increased at distances approximately 50–200 μm from the tip and, consistent with basal regions of hyphae being increasingly vacuolated, the intensity of the protein, phosphate and lipid bands decreased with distance from the tip.

The chemical composition of cell walls and cytoplasm varies dramatically in the micrometre range during the vegetative fungal life cycle of spore germination, hyphal growth and development of new spores (Jilkine et al. 2008). FTIR spectra of mature *Neurospora* spores indicated high protein, carbohydrate and nucleic acid content. The biochemical patterns seen during *Neurospora* spore development were consistent with preferential metabolite transport to sporulating structures. For example, Fig. 4.5 shows a collapsed *Rhizopus* sporangium with immature spores and synchrotron FTIR maps of the lipid, protein and sugar distribution. Despite the pronounced compositional changes seen between hyphae grown at different pH levels, the spores produced by these colonies had similar composition. Morphologically, normal hyphae grown at the same nutrient conditions did not necessarily have uniform spectra. A few hyphae had dramatically reduced carbohydrate and protein contents, which may correlate with reduced growth. Synchrotron FTIR microspectroscopy is capable of demonstrating the chemical heterogeneity between hyphae within a single colony and might contribute to our understanding of growth disparities (Jilkine et al. 2008).

Biochemical consequences of mutations are also detectable by synchrotron FTIR spectroscopy. Szeghalmi et al. (2007) compared an *A. nidulans* strain with the temperature-sensitive morphological defect *hypA1* at permissive (28 °C) and restrictive (40 °C) temperatures. The thicker hyphae of the mutant phenotype had a high content of carbohydrates, proteins and nucleic acids, even at the tip and basal regions, which lacked the typical vacuolation. Energetic-particle-irradiation is frequently employed with the aim of inducing mutations and increasing genetic variability in microorganisms or crops during the breeding process. To gain a better understanding of the interaction between energetic ions and biological systems, Liu et al. (2013) studied the biological effects of α-particle irradiation in *Rhizopus oryzae* hyphae. Synchrotron FTIR mappings revealed localised vacuolations, autolysis of the cell wall and membrane, lipid peroxidation, DNA damage and conformational changes of proteins due to ion irradiation.

Compositional changes of hyphae in response to environmental stress can also be monitored (Szeghalmi et al. 2007). Elevated pH (8.5) compared to optimal pH (6.5) reduced the carbohydrate content in the apical 200 μm, which was associated with weakened and eventually disintegrating cell walls. Surprisingly, hyphae grown at pH 5 were less robust than at pH 8.5, although fungi typically acidify their growth medium (Jilkine et al. 2008). Some fungal endophytes have been found to confer thermal tolerance to plants. For example, the ascomycete strain *Curvularia protuberata* Cp4666D was isolated from plants inhabiting geothermal soils that regularly reach 65 °C. Isenor et al. (2010) characterised hyphae of this geothermal isolate and a non-geothermal isolate from the American Type Culture Collection (CpATCC) using FT-IR-FPA imaging with the goal of comparing their biochemical composition. Crystalline mannitol was identified in some hyphae from both isolates, but more abundantly in the geothermal type. The mannitol is found at discrete, irregular locations (Isenor et al. 2010), possibly playing a role in thermal protection, not only in this fungus, but potentially in other fungal species. However, the mannitol may also be indicative of different hyphal growth phases (Isenor et al. 2010).

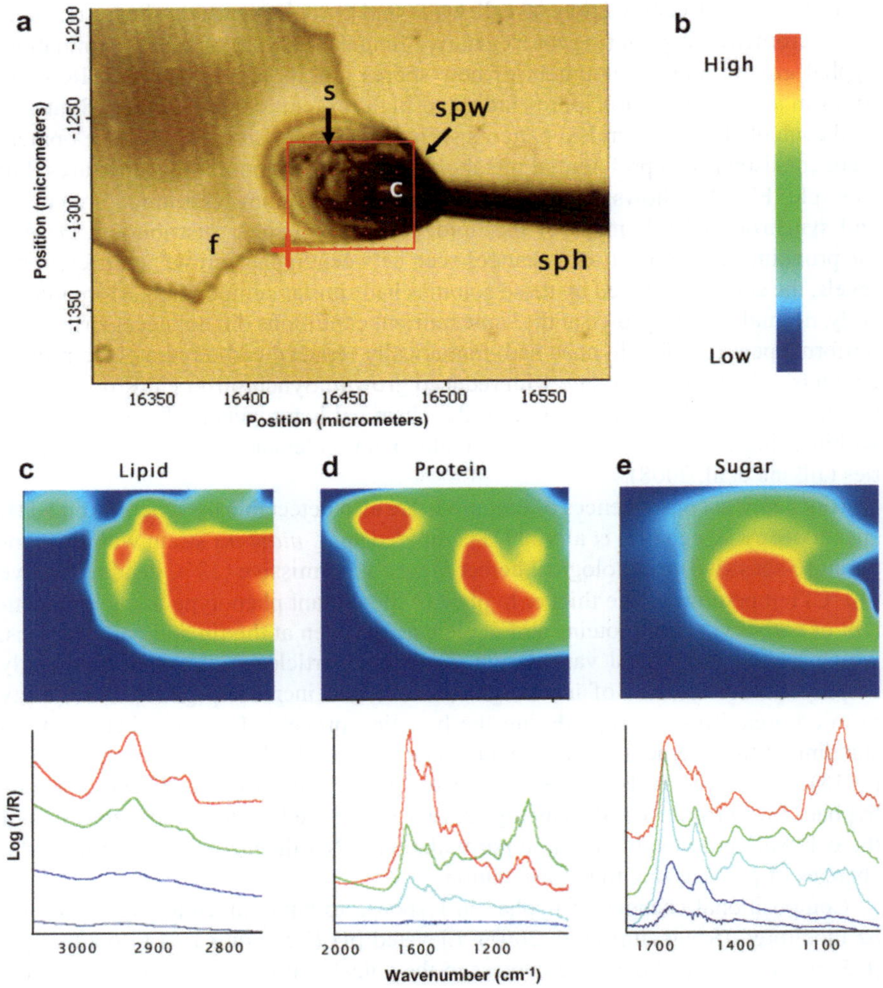

Fig. 4.5 Developing *Rhizopus* sporangium. This sporangiophore *(sph)* that supports the sporangium had collapsed onto the imaging substrate, rupturing the sporangium wall *(spw)* releasing fluids *(f)* that had bathed the developing spores, shown in the charge-coupled device (CCD) image. **a** The columella *(c)* is a dense rounded structure within the sporangium. Several oval spores *(s)* are visible within the mapped area *(red box)*. The *edges* of the box and the *points* therein indicate the centres of the pixels from which the spectra were collected. **b** The mapping colour scale. **c** Map and representative spectra for lipids: symmetric CH_2 stretch, the total area from 2864 to 2847 cm^{-1}, using 2887–2833 cm^{-1} as a baseline. **d** Map and representative spectra for proteins: amide I, the total area under 1661–1624 cm^{-1}, using 1754–831 cm^{-1} as a baseline. **e** Map and representative spectra for sugars: the total area under 1048–1013 cm^{-1}, using 1167–951 cm^{-1} as a baseline. (Reproduced with permission from Jilkine et al. 2008 © Elsevier)

The enzymatic degradation of cellophane by fungal hyphae has been investigated using FTIR microscopy with an ATR objective and FPA imaging (Oberle-Kilic et al. 2013). The decrease in peaks associated with cellophane observed adjacent to hyphae of *Armillaria mellea* and *Aspergillus niger* were consistent with their known cellulase enzyme activity (Fig. 4.6) and new protein-associated peaks adjacent to the hyphae indicated enzyme secretion. In contrast, there was no evidence of cellophane degradation in *Mucor* spp. known to have lower cellulase activity (Oberle-Kilic et al. 2013).

4.3.2 Monolayers of Single Cells, Mycelium Pellets and Fruiting Bodies

FTIR-FPA imaging was used to rapidly monitor lipid and polyunsaturated fatty acid production in the marine yeast *Rhodotorula* sp. (Vongsvivut et al. 2013). As an alternative resource to fish oil, marine microorganisms have received extensive attention for biotechnological production of polyunsaturated fatty acids for nutraceuticals. *Rhodotorula* cells were deposited on IR-reflective slides as monolayers (cyto-centrifuge Cytospin-III, Thermo Fischer Scientific, MA, USA). For semiquantitative analysis, band areas representing the proportions of unsaturated fatty acids (UFA), and total lipids were integrated under the band maxima of 3006 and 1743 cm^{-1}, respectively. A PLSR calibration analysis, with the spectral data and reference values obtained by gas chromatography (GC), was used to achieve a higher level of accuracy in determining the proportion of UFAs. The proportions of UFAs acquired from band area ratios and PLSR analysis were in good agreement with the GC results, demonstrating the potential of the PLSR approach for online quantitative monitoring of UFA production during the cell growth period (Vongsvivut et al. 2013).

Morphological growth behaviour of filamentous fungi is a key parameter in bioprocess design (Posch et al. 2013). Aiming for further insights into the formation of pellets by entanglement of *Penicillium chrysogenum* hyphae in bioreactors, microtome sections of pellets were investigated by FTIR-FPA imaging. Comparison of pellet sections from the early and late production phase revealed a reduced protein to carbohydrate content and increasing autolysis of mycelium in the pellet core (Posch et al. 2013). To establish a powerful toolbox for morphological and physiological bioprocess characterisation, these authors combined FTIR-FPA imaging with high-throughput visible light microscopy, intact cell mass spectrometry and multivariate data mining. Data from these off-line techniques were used to develop affordable in-line, real-time prediction models for fungal biomass and morphology using dielectric spectroscopy.

Fruiting bodies of the medicinal fungus *Hericium coralloides* were chemically imaged using a 16-element linear array detector (Pallua et al. 2012). Although paraffin-embedded samples were deparaffinised in octane for 8 h, no further decrease of paraffin-specific FTIR bands was observed after 2 h dewaxing. FTIR imaging

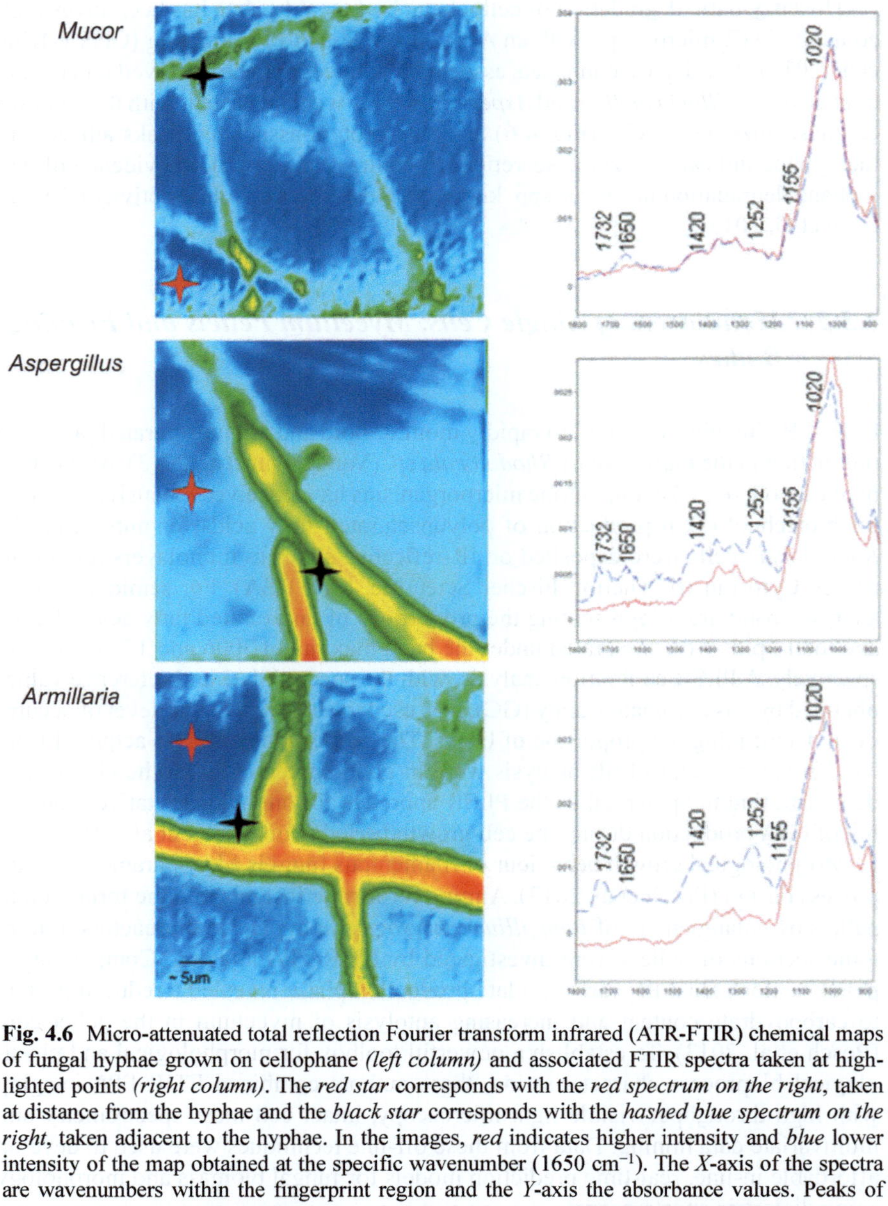

Fig. 4.6 Micro-attenuated total reflection Fourier transform infrared (ATR-FTIR) chemical maps of fungal hyphae grown on cellophane *(left column)* and associated FTIR spectra taken at highlighted points *(right column)*. The *red star* corresponds with the *red spectrum on the right*, taken at distance from the hyphae and the *black star* corresponds with the *hashed blue spectrum on the right*, taken adjacent to the hyphae. In the images, *red* indicates higher intensity and *blue* lower intensity of the map obtained at the specific wavenumber (1650 cm^{-1}). The *X*-axis of the spectra are wavenumbers within the fingerprint region and the *Y*-axis the absorbance values. Peaks of interest that changed in intensity, appeared or diminished are labelled. *Italicised labels* are associated with changes in carbohydrate chemistry (decomposition of cellophane). Remaining peaks are associated with the appearance of proteins (enzymes) in the sample. (Reproduced with permission from Oberle-Kilic et al. 2013 © Mycological Society of China)

clearly indicated enhanced protein content in the hymenial tissue formed by basidia, basidioles and cystidia compared to the trama tissue of sterile hyphae. Hierarchical cluster analysis produced a cluster structure clearly representing the histological features of trama tissue, hymenium tissue and accumulations of basidiospores based on chemical differences (Pallua et al. 2012). Thus, many features of the tissue that require staining for light microscopy histology can be highlighted with false colour representation of FTIR data from unstained tissue.

4.4 Fungal Discrimination

In this section, spectroscopic results are first presented to lay the basis for the following FTIR microscopic analyses and illustrate the potential for future investigations. An intelligent combination of high-throughput spectroscopic analyses with selected imaging will likely solve many scientific questions more efficiently than imaging alone. The possibility of measuring the chemical composition over such a wide range of spatial resolution by FTIR spectroscopy and microscopy is a great advantage of FTIR analysis. This enables the use of variable sample sizes and aids in taking representative subsamples of each class, which are essential for class discrimination (e.g. fungal taxa). The relationship between intra- and inter-class variance is decisive for class discrimination where smaller differences between class mean spectra require reduced intra-class variance. Rigorously standardised cultivation and measurement procedures are usually necessary to discriminate many closely related strains within a species. In contrast, few classes with samples from various environments can be discriminated, provided that class differences are large enough. The extent of chemotaxonomic differences within and between taxonomic classes at comparable levels can vary considerably (Naumann 2009). An overview of FTIR analyses for the identification of filamentous fungi and yeasts is given in Santos et al. (2010).

4.4.1 Spectroscopic Techniques

For classifying and identifying microorganisms, FTIR analysis combined with multivariate analysis offers valuable opportunities (Naumann et al. 1991). Using FTIR spectroscopy, spectra can be recorded for microorganism solutions dried to transparent films or transferred from agar plates with a velvet stamp (Helm et al. 1991; Naumann 2000; Ngo Thi et al. 2000). These methods, originally developed for the classification of bacteria, were transferable to unicellular fungal cells such as food-borne and pathogenic yeasts (Kümmerle et al. 1998; Timmins et al. 1998; Mounier et al. 2006; Toubas et al. 2007) and spores (Fischer et al. 2006). Compared to macrospectroscopy, FTIR micro(spectro)scopy of yeast microcolonies vastly reduced the time of cultivation and identification from days to several hours (Wenning et al. 2002; Ngo Thi et al. 2003).

Obtaining reproducible spectra of mycelium required substantial methodological adaptations. Transparent films of microorganisms contain many individuals, which can easily be a representative subsample of a whole colony. Mycelium belonging to a single individual can grow several centimetres in a few days and show considerable chemical differences related to parameters such as hyphal age. Therefore, taking a representative subsample for one or numerous individuals and preparing a homogeneous film is a challenge. Powdering of mycelia sampled from a whole agar plate resulted in reproducible spectra for FTIR-ATR analysis. Aiming to discriminate 26 strains from 24 wood-destroying fungal species, cluster analysis classified 99 % of 233 spectra correctly on the species level according to their spectral similarity (Naumann 2009). This result demonstrates the excellent potential of this method to discriminate fungal strains grown in pure culture under standardised culture conditions. For example, differences between FTIR spectra of six strains from six phylogenetic orders are shown in Fig. 4.7.

A number of publications explored the classification of soil-borne and phytopathogenic fungal strains, especially of *Fusarium* but also of *Colletotrichum, Pythium, Rhizoctonia, Verticillum* etc. (e.g. Nie et al. 2007; Linker and Tsor 2008; Salman et al. 2011, 2012). Direct FTIR-ATR analysis of wet hyphae can produce very rapid results for genus and strain classification (Linker and Tsror 2008) but with a slightly lower percentage of correct identifications compared to dried samples. In comparison to the ATR method, pressing powdered dried fungal cell mass and KBr into pellets (Nie et al. 2007) is a laborious procedure.

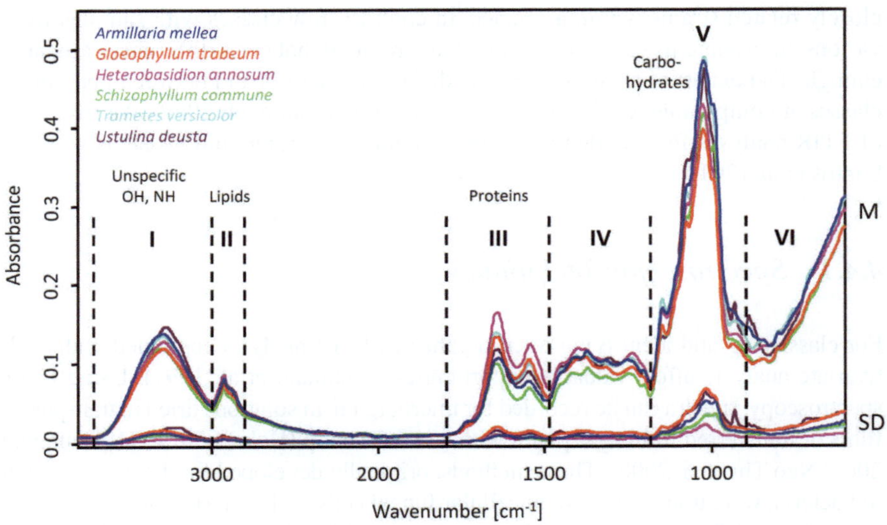

Fig. 4.7 Attenuated total reflection Fourier transform infrared (ATR-FTIR) spectra of mycelium from 6 fungal strains. Mean spectra of mycelium grown on 9 agar plates per strain from 3 independent experimental sets. Ranges *I–VI* were used for cluster analysis. *Above*, dominating functional groups or macromolecules are specified. (Adapted with permission from Naumann 2009 © The Royal Society of Chemistry)

Analysis of root tips colonised with distinct ectomycorrhizal fungal species collected *in situ* in the forest demonstrated that not only pure fungal cultures enable species discrimination but also environmental samples from various tree species and geographic areas showed species-specific spectral features (Pena et al. 2014). By means of FTIR-ATR analysis and hierarchical cluster analysis, 97% of 181 sample spectra were correctly assigned to their species *(Amanita rubescens, Cenococcum geophilum, Lactarius subdulcis, Russula ochroleuca, Xerocomus pruinatus)*. Using 73 *Pleurotus* strains, it was shown that spectral data could be exploited for specimen identification with the aid of a dichotomous identification key (Zervakis et al. 2012). Furthermore, the correct classification of 11 lichen species belonging to the families Physciaceae and Parmeliaceae demonstrated the potential of FTIR spectroscopy to distinguish lichens, a symbiosis of a fungus and a photosynthetic partner (Alcantara et al. 2007).

To differentiate filamentous fungi from food spoilage, Shapaval et al. (2012) developed a high-throughput microcultivation protocol in a microtiter plate system using 350 μl of liquid culture medium in each of the 200 wells and 5 days of cultivation. Mycelia from multiple independent cultivations of 59 strains belonging to 10 genera and 19 species were analysed by FTIR spectroscopy. Artificial neural network analyses for classification and validation resulted in 94% correctly identified spectra, both at the species and genus level. The percentage of correct identification varied considerably among species (75–100%). Lecellier et al. reduced the cultivation time for mycelium production in liquid medium to 48 h after spore inoculation (2014) and increased the number of fungal strains to 486 representing 43 genera and 140 species (2015). Their IR-spectral database, built with 288 strains and based on partial least squares discriminant analysis (PLS-DA), assigned 99.17 and 92.3% of 105 different strains correctly at the genus and species levels, respectively.

In biomedicine, the potential of FTIR spectroscopy to discriminate dermatophytes causing skin infections, for example *Trichophyton* species, was evaluated by Salem et al. (2010) and Ergin et al. (2013).

Direct detection in and discrimination of fungi from other material is desirable to preclude laborious and time-consuming isolation of a pure culture or for fungi which do not grow in culture. As proof of principle, Irudayaraj et al. (2002) showed FTIR-photoacoustic spectroscopy to successfully detect and differentiate various bacteria and fungi (*Saccharomyces cerevisiae, A. niger* and *Fusarium verticillioides*) smeared on apple skin.

4.4.2 Microscopic Techniques

FTIR microscopy is an excellent tool to detect fungi within other materials. As an example, FPA imaging enabled visualising the distribution of fungal mycelium within a beech wood section (Naumann et al. 2005). The three classes, determined by cluster analysis of the 4096 FTIR spectra from the FPA data set, corresponded very well with the distribution of fungal mycelium, wood and air in the vessels visible in the microscopic image (Fig. 4.8). In addition to discrimination of the fungal mycelium from other material such as wood and air, FTIR microscopy can

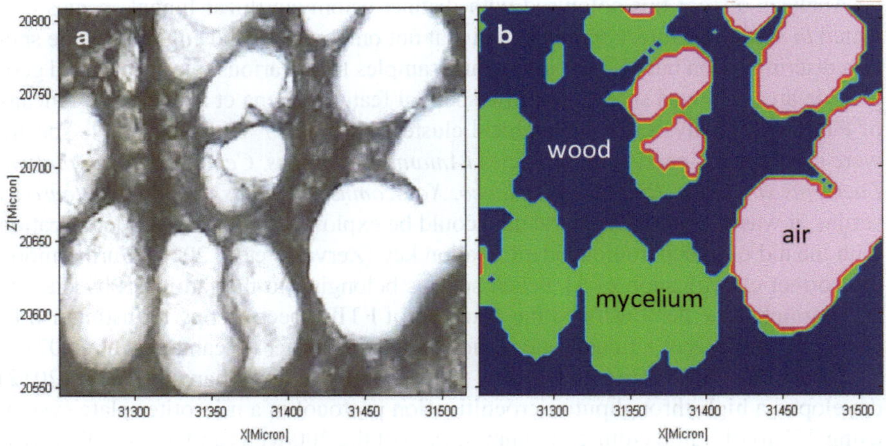

Fig. 4.8 Light microscopic image of fungal mycelium in a beech wood section (**a**) and cluster analysis of Fourier transform infrared (FTIR) spectra from a focal plane array (FPA) data set in the range 1188–922 cm^{-1} (**b**). (Adapted with permission from Naumann et al. 2005 © Elsevier and reproduced with permission from Schulz et al. 2014 © Wiley-VCH Verlag GmbH & Co. KGaA)

differentiate between mycelium of diverse species. For this purpose, spectra can be either extracted from an FPA data set or recorded by repeated measurements with a single-channel detector, as in the following example. Mycelia of the wood-destroying fungi *Trametes versicolor* and *Schizophyllum commune* within vessels were successfully discriminated from each other as well as from wood and air spectra (Naumann et al. 2005, Fig. 4.9). Interestingly, the chemical composition of mycelium within vessels and grown on top of wood blocks differed. Furthermore, transmission FTIR microscopy of thin surface scratches of potatoes discriminated uninfected samples and regions infected with the fungal pathogens *Colletotrichum, Rhizoctonia* and *Helminthosporium* (Erukhimovitch et al. 2010).

As previously mentioned in relation to fatty acid quantification, cellular monolayers on IR-reflective slides imaged by FPA can be used to discriminate marine yeast strains. Using independent validation of models established by PLS-DA (Vongsvivut et al. 2013), correctly differentiated 100 % of the test samples from four *Rhodotorula* strains and one *Thraustochytrium* strain.

4.5 Biotic Interactions

This section will review the applications of FTIR to studying the fungal degradation of biotic materials from once-living organisms, as well as interactions of fungi with other living organisms. Understanding the diversified degradation processes of complex dead biological material (e.g. wood) and detecting fungal mycelium within a substrate still pose enormous challenges, but studying interactions between living organisms multiplies that complexity. Not only do interaction partners need

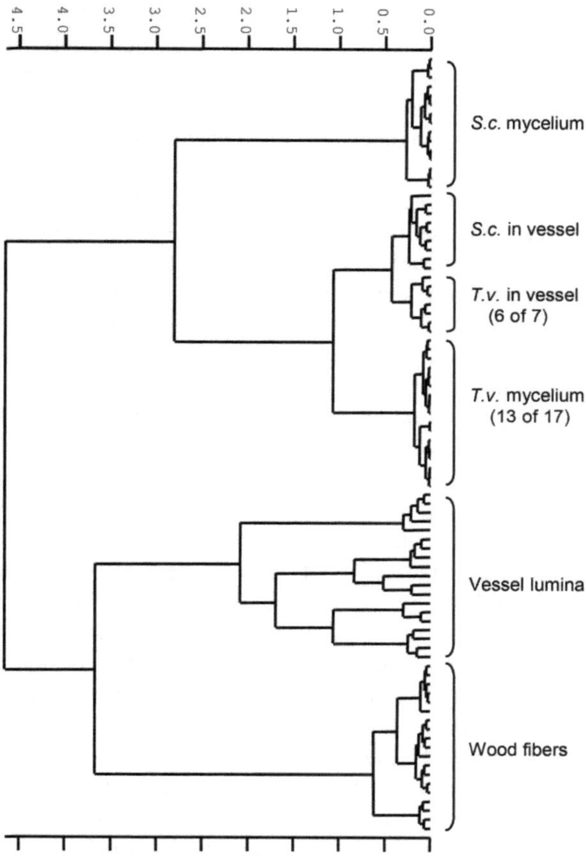

Fig. 4.9 Cluster analysis of vector-normalised spectra obtained from beech wood fibres (wood), mycelium of *Schizophyllum commune (S. c.)* and *Trametes versicolor (T. v.)* growing on the wood surface and in vessel lumina *(S. c. in vessel, T. v. in vessel)* and vessel lumina without mycelium *(vessel lumina)*. (Reproduced with permission from Naumann et al. 2005, © Elsevier)

to be detected within a complex three-dimensional matrix of an interaction partner, but such interactions are often accompanied by chemical changes related to both interacting partners over time. For instance, in plant–fungus interactions, the fungal partner produces degrading enzymes and uses plant metabolites to synthesise its new materials, while the plant composition can change as a function of defence mechanisms and fungal decay.

4.5.1 Wood Degradation

FTIR spectroscopy of decayed wood samples has provided valuable insights into wood degradation (Pandey and Pitman 2003; Fackler and Schwanninger 2012 and references therein). Understanding fungal degradation of trees and construction wood is of high priority because collapse poses a serious safety risk. Early on in fungal decay, mechanical strength can be dramatically reduced long before the infection becomes visible at the wood surface (Fackler et al. 2010). Traditional wet chemical analyses often fail to detect the small chemical changes during the beginning (incipient) decay stages. Several higher fungi are capable of degrading wood, a highly durable structure of cellulose fibrils, hemicelluloses and lignin, associated with three main decay mechanisms. Primary degradation of lignin is referred to as white rot and increases the relative cellulose content (Schwarze 2007; Fackler et al. 2011), while brown rot fungi degrade mainly cellulose fibres, leaving the brownish lignin (Schwarze 2007; Fackler et al. 2010). Soft rot in damp environments decays lignin and forms cellulose cavities next to the hyphae (Schwarze 2007). Transitions between these decay mechanisms are possible (Schwarze 2007).

Degradation of poplar wood by the multicomponent enzyme Onozuka RS derived from the fungus *Trichoderma viride* was studied by Gierlinger et al. (2008). In a custom-built fluidic cell, wood cross sections were treated using a constant flow of Onozuka RS enzyme solution with cellulase and other cell-wall-degrading activities. This method allows the degradation process to be monitored at a distinct sample position in the wet state using FTIR microspectroscopy, while flow ensures uniform accessibility of the enzyme and excludes feedback inhibition by end products. The fast, selected hydrolysis of the cellulose-rich gelatinous (G) layer was demonstrated in this study. A high proportion of G-layer, a desirable resource for bioethanol production, could be achieved by wind exposure or selection of species provenances (Gierlinger et al. 2008).

To image early brown and white rot, Fackler et al. (2010, 2011) employed FTIR imaging with a 2×8 linear array detector having a potential spatial resolution of 6.25×6.25 µm. In sections of spruce wood, PCA-based image analysis was used to spatially resolve the degree of brown rot and a model was built using PLS-DA to detect incipient brown rot (Fackler et al. 2010, Fig. 4.10). As highly indicative for incipient brown and white rot, there was a significant reduction of glycosidic linkages within polysaccharides at 1160 cm^{-1} in relation to carbohydrate ring structures. Fungal decay of wood is reduced at low moisture content. In wood-plastic composites (WPC), embedding wood particles in a polymer matrix decreases moisture uptake so effectively that product testing requires sensitive test methods. FTIR-ATR spectroscopy of the WPC surface demonstrated differences in wood decay by a white rot fungus, a brown rot fungus and a surface mould at an early stage as well as effects of weathering and nitrogen-containing flame retardants (Naumann et al. 2012a[1], 2012b[1]).

Decay of wood by the fungus *Hypholoma fasciculare* in interaction with bacteria has been examined by Weißhaupt et al. (2013[1]) by FTIR spectroscopy, showing that fungal decomposition of wood blocks was similar with and without bacterial coculture. Hence, the bacteria coexisting with the fungus seemed to compete for the same nitrogen sources and not support fungal nitrogen nutrition by diazotrophic activity. Aside from studying wood degradation, FTIR analyses can provide valuable

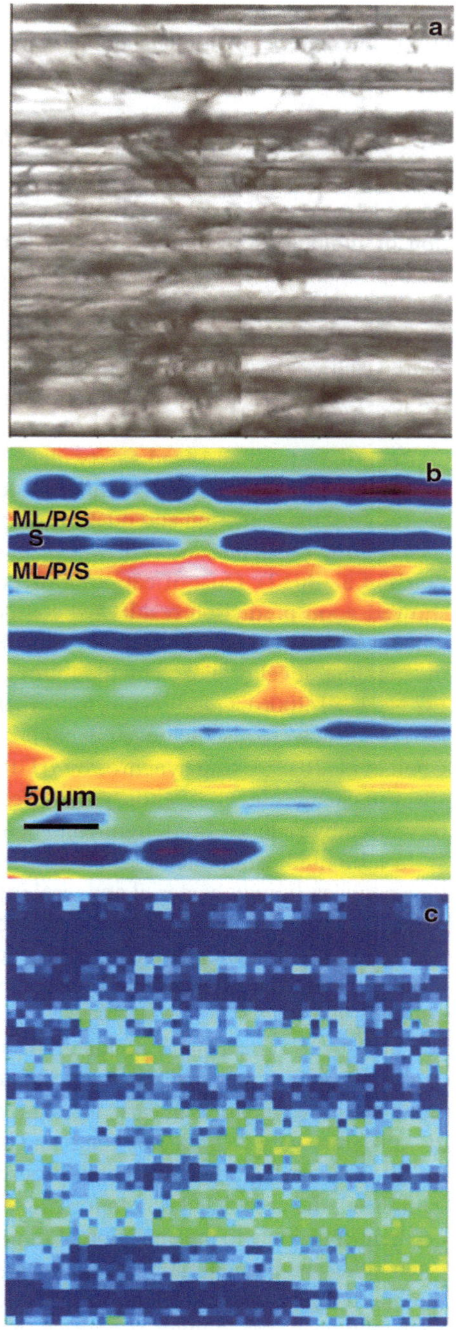

Fig. 4.10 Localisation of early brown rot degradation within a radial thin section of spruce wood with Fourier transform infrared (FTIR) imaging microspectroscopy: **a** charge-coupled device (CCD) camera image of 10 tracheids of a degraded spruce wood section from a sample degraded by the brown rot fungus *Gloeophyllum trabeum* for 4 weeks. **b** FTIR pseudo-colour spectral absorbance image. Zones of high total absorbance *(red)* show a high contribution of middle lamella (ML) and primary cell walls (P) and outer layers of secondary cell walls S1, named as *ML/P/S*. Those with low total absorbance *(blue)* can be assigned to regions with a high contribution of secondary cell walls S2 and pits, named *S*. **c** Example of a partial least squares discriminant analysis image gained through multivariate image analysis in which degraded pixels are plotted *yellow* to *light green* and are found in *ML/P/S* regions of the section in 6.25-µm-pixel resolution. (Reproduced with permission from Fackler and Schwanninger 2012 © Springer)

insights into the whole decomposition process from forest litter to organic soil layers (Haberhauer et al. 1998).

4.5.2 Detection of Fungal Infestation and Mycotoxins

Early detection of fungal infections is crucial in agriculture to minimise the application of fungicides and harvest losses. FTIR-photoacoustic spectroscopy detected soybean rust in soy leaves by *Phakopsora pachyrhizi* earlier than the conventional visual method (Andrade et al. 2008). Furthermore, black dot disease caused by *Colletotrichum coccodes* in potato was detected before visual morphological signs appeared and epidermal tissue scratched from uninfected and infected potatoes was correctly classified using FTIR microscopy in transmission mode (Erukhimovitch et al. 2007).

Regarding food and feed safety, detection of fungal infections and mycotoxins are also an important issue. A number of studies demonstrated the suitability of FTIR analysis for the detection and quantification of fungal contaminations in grains (e.g. *Aspergillus, Fusarium* in corn, wheat: Gordon et al. 1993, 1997, Kos et al. 2002, Peiris et al. 2012). Quantification of mycotoxins using reference values and PLSR showed good prediction results for compounds such as deoxynivalenol (DON) in wheat (Abramović et al. 2007) and corn (Kos et al. 2003) in ranges associated with the maximal level proposed by the European Commission (Commission Regulation EC No 1881/2006). Good results for screening mycotoxin (DON, aflatoxin) content in feedstuffs were also achieved using near-infrared (NIR) analysis, as reviewed by Cheli et al. 2012. Using synchrotron FTIR microscopy, Singh et al. (2011) revealed spectral changes in wheat endosperm infected with *Aspergillus glaucus*, proposed to be caused by reduced lipid, lignin and cellulose content.

In a biomedical application, FTIR microscopy was able to detect and discriminate between fungal, bacterial and viral infections in cell cultures of monkey kidney cells (Erukhimovitch et al. 2013). In addition, FTIR-photoacoustic spectroscopy was used to follow chemical changes, which are linked with morphological changes in lung tissues during infection by the fungus *Paracoccidioides brasiliensis* (Morato et al. 2013).

During the time course of material colonisation by fungi, there are chemical changes from an increasing amount of growing fungal biomass and secreted fungal enzymes, especially during early infestation stages. In later infestation stages, material is progressively degraded by fungi and in the end, fungal biomass may be resorbed by autolysis or degraded by other microorganisms. In cases where living biomaterial is infested, the associated resistance reactions can result in chemical changes, which can help contribute to the detection of fungal infestations. In such studies, the focus is often either the biomaterial degradation or fungal development, without considering the interaction partner. For simple detection of fungal infestations, determining the contributions of the individual interaction partners is not essential. However, a better understanding of such interactions requires a detailed look at each of the interaction partners, examples of which appear in the following section.

4.5.3 Interaction with Living Organisms

For evaluating the resistance of elms *(Ulmus minor)* to the Dutch elm disease-causing fungus *Ophiostoma novo-ulmi*, Martín et al. (2005) studied chemical changes 40 days following elm seedling inoculation using FTIR microscopy. Spectra of transverse sections from resistant trees differed from those of susceptible and control trees, indicating increased lignin and suberin formation. Comparison of *U. minor* with the often less susceptible hybrid *U. minor* × *Ulmus pumila* showed increased lignin levels in the xylem tissues of the hybrid, suggested to result from an earlier response to the infection (Martín et al. 2007). Moreover, study of the resistance of *Arabidopsis thaliana* mutants against powdery mildew caused by the fungus *Erysiphe cichoracearum* (Raab and Vogel 2004) revealed two mutants with novel forms of resistance different from known defence pathways. Synchrotron FTIR microscopy of leaf preparations cleared of pigments and waxes indicated similar modifications of the cell wall's pectin composition in both resistant mutants (Raab and Vogel 2004). Metabolomic approaches were used to study interactions of the fungus *Magnaporthe grisea*, the cause of rice blast disease, with the grass *Brachypodium distachyon* (Allwood et al. 2006). Susceptible and resistant accessions of *B. distachyon* were shown to differ in their fatty acid composition using FTIR spectroscopy. Subsequent electrospray ionisation mass spectrometry-mass spectrometry (ESI-MS-MS) identified several phospholipids and suggested differential phospholipid processing of membrane lipids due to resistance mechanisms and development of disease symptoms (Allwood et al. 2006). Resistance of *A. thaliana* against the necrotrophic fungus *Botrytis cinerea*, the cause of grey mould, is conferred by ethylene signalling, prompting comparison of metabolic responses from wild-type and ethylene signalling mutants (Lloyd et al. 2011). FTIR-FPA imaging correlated spatial accumulations of hydroxycinnamates and monolignols near infection sites of *B. cinerea* with ethylene-mediated resistance of the mutants and thus helped to reveal the contribution of cell wall modifications by hydroxycinnamates and monolignols to the defence against *B. cinerea*.

4.6 Conclusions and Outlook

A better understanding of the complex metabolic processes in fungi, the chemotaxonomic characterisation of fungi and their abiotic and biotic interactions requires rapid, versatile and cost-effective analysis methods. In this respect, FTIR microscopy provides a competitive metabolic fingerprinting method of main components such as lipids, proteins, nucleic acids and carbohydrates, usually down to the mg/g range. Because FTIR analyses require no or minimal sample preparation, associated artefacts are minimised. Particularly, the gain in analysis speed compared to methods such as genomics and mass-spectrometric metabolomics is significant. Another great advantage is the combination of FTIR microscopy and spectroscopy, enabling an extremely wide range of sample types, sizes and consequently considerable spatial resolution. Moreover, the lack of chemical consumables makes FTIR analyses exceptionally cost-efficient and environmentally friendly. All of these factors contribute to making FTIR analyses an effective and efficient tool for screening numerous complex samples for unforeseen metabolic differences. The potential of FTIR analyses are substantially enhanced when metabolic differences are further analysed using GC- or LC-MS-based metabolomics to elucidate the exact molecular changes (Allwood et al. 2008). Such an approach enables more detailed spectral band assignment to chemical components for future FTIR analyses.

FTIR analyses are capable of advancing our understanding of fungal biology, especially in areas which can exploit the capability of rapidly monitoring complex chemical changes associated with processes such as growth and development during the fungal life cycle. FTIR imaging enables comparison of chemical images at progressing developmental stages, which includes spores, hyphae, mycelia, and fruiting bodies.

A special challenge in fungal biology is assessing the immense chemotaxonomic diversity of fungi. FTIR analyses provide a valuable tool to screen fungal isolates prior to genetic or mass-spectrometric methods. Due to the superior analysis speed and large range of spatial resolution, FTIR analyses can be used to classify chemically similar isolates, reducing the number of more laborious and expensive measurements. The screening of environmental samples (e.g. mycorrhised root tips) offers insight into inter- versus the intra-specific biochemical diversity.

In particular, the study of fungus–host interactions is a complex and interesting field. The investigation of each individual interaction partner should enable the detection of chemical features typical during interaction. Valuable new insights are especially expected from an interdisciplinary approach involving mycologists, botanists, medical scientists, biochemists, statisticians, molecular biologists and instrument manufacturers.

Acknowledgements AN thanks the reviewer Susan Kaminskyj, the editor Tanya Dahms, Susanne Wurst, and Hartwig Schulz for critical reading of the manuscript and valuable comments that improved the manuscript. Work related to publications labelled with superscript [1] was conducted by AN during 2010–2012 at BAM Federal Institute for Materials Research and Testing in division 4.1, Biology in Materials Protection and Environmental Issues, and division 6.3, Durability of Polymers.

References

Abramović B, Jajić I, Abramović B, Ćosić J, Jurić V (2007) Detection of deoxynivalenol in wheat by Fourier transform infrared spectroscopy. Acta Chim Slov 54:859–867

Alcantara GB, Honda NK, Ferreira MMC, Ferreira AG (2007) Chemometric analysis applied in 1 H HR-MAS NMR and FT-IR data for chemotaxonomic distinction of intact lichen samples. Anal Chim Acta 595:3–8

Allwood JW, Ellis DI, Goodacre R (2008) Metabolomic technologies and their application to the study of plants and plant-host interactions. Physiol Plant 132:117–135

Allwood JW, Ellis DI, Heald JK, Goodacre R, Mur LAJ (2006) Metabolomic approaches reveal that phosphatidic and phosphatidyl glycerol phospholipids are major discriminatory non-polar metabolites in responses by Brachypodium distachyon to challenge by Magnaporthe grisea. Plant J 46:351–368

Andrade LHC, Freitas PG, Mantovani BG, Figueiredo MS, Lima RA, Lima SM et al (2008) Detection of soybean rust contamination in soy leaves by FTIR photoacoustic spectroscopy. Eur Phys J Spec Top 153:539–541

Chalmers JM, Griffiths PR (2002) Handbook of vibrational spectroscopy, vol 1–5. Wiley, Chichester

Cheli F, Battaglia D, Pinotti L, Baldi A (2012) State of the art in feedstuff analysis: a technique-oriented perspective. J Agric Food Chem 60:9529–9542

Commission Regulation (EC) No (1881/2006 of 19 December 2006). Setting maximum levels for certain contaminants in foodstuffs. Off J Eur Union L364:5–24

Ergin Ç, İlkit M, Gök Y, Özel MZ, Çon AH, Kabay N et al (2013) Fourier transform infrared spectral evaluation for the differentiation of clinically relevant Trichophyton species. J Microbiol Methods 93:218–223

Erukhimovitch V, Tsror (Lahkim) L, Hazanovsky M, Talyshinsky M, Souprun Y, Huleihel M (2007) Early and rapid detection of potato's fungal infection by Fourier transform infrared microscopy. Appl Spectrosc 61:1052–1056

Erukhimovitch V, Tsror (Lahkim) L, Hazanovsky M, Huleihel M (2010) Direct identification of potato's fungal phyto-pathogens by Fourier-transform infrared (FTIR) microscopy. Spectroscopy 24:609–619

Erukhimovitch V, Huleihil M, Huleihel M (2013) Identification of contaminated cells with viruses, bacteria, or fungi by Fourier transform infrared microspectroscopy. J Spectrosc 317458:1–6

Fackler K, Schwanninger M (2012) How spectroscopy and microspectroscopy of degraded wood contribute to understand fungal wood decay. Appl Microbiol Biotechnol 96:587–599

Fackler K, Thygesen LG (2013) Microspectroscopy as applied to the study of wood molecular structure. Wood Sci Technol 47:203–222

Fackler K, Stevanic JS, Ters T, Hinterstoisser B, Schwanninger M, Salmén L (2010) Localisation and characterisation of incipient brown-rot decay within spruce wood cell walls using FT-IR imaging microscopy. Enzyme Microb Technol 47:257–267

Fackler K, Stevanic JS, Ters T, Hinterstoisser B, Schwanninger M, Salmén L (2011) FT-IR imaging microscopy to localise and characterise simultaneous and selective white-rot decay within spruce wood cells. Holzforschung 65:411–420

Fischer G, Braun S, Thissen R, Dott W (2006) FT-IR spectroscopy as a tool for rapid identification and intra-species characterization of airborne filamentous fungi. J Microbiol Methods 64:63–77

Gierlinger N, Goswami L, Schmidt M, Burgert I, Coutand C, Rogge T, Schwanninger M (2008) In situ FT-IR microscopic study on enzymatic treatment of poplar wood cross-sections. Biomacromolecules 9:2194–2201

Gordon SH, Green RV, Wheeler BC, James C (1993) Multivariate FTIR analysis of substrates for protein, polysaccharide, lipid and microbe content: potential for solid-state fermentations. Biotechnol Adv 11:665–675

Gordon SH, Schudy RB, Wheeler BC, Wicklow DT, Greene RV (1997) Identification of Fourier transform infrared photoacoustic spectral features for detection of *Aspergillus flavus* infection in corn. Int J Food Microbiol 35:179–186

Günzler G, Gremlich H-U (2002) IR spectroscopy. Wiley-VCH, Weinheim

Haberhauer G, Rafferty B, Strebl F, Gerzabek MH (1998) Comparison of the composition of forest soil litter derived from three different sites at various decompositional stages using FTIR spectroscopy. Geoderma 83:331–342

Helm D, Labischinski H, Naumann D (1991) Elaboration of a procedure for identification of bacteria using Fourier-transform IR spectral libraries: a stepwise correlation approach. J Microbiol Methods 14:127–142

Irudayaraj J, Yang H, Sakhamuri S (2002) Differentiation and detection of microorganisms using Fourier transform infrared photoacoustic spectroscopy. J Mol Struct 606:181–188

Isenor M, Kaminskyj SGW, Rodriguez RJ, Redmance RS, Gough KM (2010) Characterization of mannitol in Curvularia protuberata hyphae by FTIR and Raman spectromicroscopy. Analyst 135:3249–3254

Jilkine K, Gough KM, Julian R, Kaminskyj SGW (2008) A sensitive method for examining whole-cell biochemical composition in single cells of filamentous fungi using synchrotron FTIR spectromicroscopy. J Inorg Biochem 102:540–546

Kaminskyj S, Jilkine K, Szeghalmi A, Gough K (2008) High spatial resolution analysis of fungal cell biochemistry—bridging the analytical gap using synchrotron FTIR spectromicroscopy. FEMS Microbiol Lett 284:1–8

Kazarian SG, Chan KLA (2010) Micro- and macro-attenuated total reflection Fourier transform infrared spectroscopic imaging. Appl Spectrosc 64:135–152

Kos G, Lohninger H, Krska R (2002) Fourier transform infrared spectroscopy with attenuated total reflection (FT-IR/ATR) as a tool for the detection of Fusarium fungi on maize. Vib Spectrosc 29:115–119

Kos G, Lohninger H, Krska R (2003) Validation of chemometric models for the determination of deoxynivalenol on maize by mid-infrared spectroscopy. Mycotoxin Res 19:149–153

Krafft C, Steiner G, Beleites C, Salzer R (2009) Disease recognition by infrared and Raman spectroscopy. J Biophoton 2:13–28

Kümmerle M, Scherer S, Seiler H (1998) Rapid and reliable identification of food-borne yeasts by Fourier-transform infrared spectroscopy. Appl Environ Microbiol 64:2207–2214

Lecellier A, Mounier J, Gaydou V, Castrec L, Barbier G, Ablain W, Manfait M, Toubas D, Sockalingum GD (2014) Differentiation and identification of filamentous fungi by high-throughput FTIR spectroscopic analysis of mycelia. Int J Food Microbiol 168–169:32–41

Lecellier A, Gaydou V, Mounier J, Hermet A, Castrec L, Barbier G, Ablain W, Manfait M, Toubas D, Sockalingum GD (2015) Implementation of an FTIR spectral library of 486 filamentous fungi strains for rapid identification of molds. Food Microbiol 45:126–143

Linker R, Tsror (Lahkim) L (2008) Discrimination of soil-borne fungi using Fourier transform infrared attenuated total reflection spectroscopy. Appl Spectrosc 62:302–305

Liu J, Qi Z, Huang Q, Wei X, Ke Z, Fang Y, Tian Y, Yu Z (2013) Study of energetic-particle-irradiation induced biological effect on Rhizopus oryzae through synchrotron-FTIR micro-spectroscopy. J Mol Struct 1031:1–8

Lloyd AJ, Allwood WJ, Winder CL, Dunn WB, Heald JK, Cristescu SM, Sivakumaran A, Harren FJM, Mulema J, Denby K, Goodacre R, Smith AR, Mur LAJ (2011) Metabolomic approaches reveal that cell wall modifications play a major role in ethylene-mediated resistance against *Botrytis cinerea*. Plant J 67:852–868

Lyng F, Gazi E, Gardner P (2011) Preparation of tissues and cells for infrared and Raman spectroscopy and imaging. In: Moss D (ed) Biomedical applications of synchrotron infrared microspectroscopy. Royal Society of Chemistry, Cambridge, p 145–189

Mariey L, Signolle JP, Amiel C, Travert J (2001) Discrimination, classification, identification of microorganisms using FTIR spectroscopy and chemometrics. Vib Spectrosc 26:151–159

Martín JA, Solla A, Woodward S, Gil L (2005) Fourier transform infrared spectroscopy as a new method for evaluating host resistance in Dutch elm disease complex. Tree Physiol 25:1331–1338

Martín JA, Solla A, Woodward S, Gil L (2007) Detection of differential changes in lignin composition of elm tissues inoculated with Ophiostoma novo-ulmi using Fourier transform infrared spectroscopy. For Pathol 37:187–191

Michell AJ, Higgins HG (2002) Infrared spectroscopy in Australian forest products research. CSIRO forestry and forestry products, Melbourne

Mohaček-Grošev V, Božac R, Puppels GJ (2001) Vibrational spectroscopic characterization of wild growing mushrooms and toadstools. Spectrochim Acta A 57:2815–2829

Morato EM, Morais GR, Sato F, Medina AN, Svidinski TI, Baesso ML et al (2013) Morphological and structural changes in lung tissue infected by *Paracoccidioides brasiliensis*: FTIR photoacoustic spectroscopy and histological analysis. Photochem Photobiol 89:1170–1175

Mounier J, Goerges S, Gelsomino R, Vancanneyt M, Vandemeulebroecke K, Hoste B, Brennan NM, Scherer S, Swings J, Fitzgerald GF, Cogan TM (2006) Sources of the adventitious microflora of a smear-ripened cheese. J Appl Microbiol 101:668–681

Movasaghi Z, Rehman S, Rehman I (2008) Fourier transform infrared (FTIR) spectroscopy of biological tissues. Appl Spectrosc Rev 43:134–179

Naumann D (2000) Infrared spectroscopy in microbioloy. In: Meyers RA (ed) Encyclopedia of analytical chemistry. Wiley, Chichester, pp 102–131

Naumann A (2009) A novel procedure for strain classification of fungal mycelium by cluster analysis and artificial neural network analysis of Fourier transform infrared (FTIR) spectra. Analyst 134:1215–1223

Naumann D, Helm D, Labischinski H (1991) Microbiological characterizations by FT-IR spectroscopy. Nature 351:81–82

Naumann A, Navarro-González M, Peddireddi S, Kües U, Polle A (2005) Fourier transform infrared microscopy and imaging: detection of fungi in wood. Fungal Genet Biol 42:829–835

Naumann A, Peddireddi S, Kües U, Polle A (2007) Fourier transform infrared microscopy in wood analysis. In: Kües U (ed) Wood production, wood technology and biotechnological impacts. Universitätsverlag Göttingen, Göttingen, pp 179–196

Naumann A, Stephan I, Noll M (2012a) Material resistance of weathered wood-plastic composites against fungal decay. Int Biodeterior Biodegradation 75:28–35

Naumann A, Seefeldt H, Stephan I, Braun U, Noll M (2012b) Material degradation of flame retarded wood-plastic composites against fire and fungal decay. Polymer Degrad Stab 97:1189–1196

Ngo Thi NA, Kirschner C, Naumann D (2000) FT-IR microspectrometry: a new tool for characterizing micro-organisms. Proc SPIE 3918:36–44

Ngo Thi NA, Kirschner C, Naumann D (2003) Characterization and identification of microorganisms by FT-IR microspectrometry. J Mol Struct 661–2:371–380

Nie M, Zhang WQ, Xiao M, Luo JL, Bao K, Chen JK, Li B (2007) FT-IR spectroscopy and artificial neural network identification of Fusarium species. J Phytopathol 155:364–367

NIST Chemistry WebBook (2014) NIST Chemistry WebBook. National Institute of Standards and Technology, Gaithersburg. http://webbook.nist.gov/chemistry/. Accessed 10 April 2014

Oberle-Kilic J, Dighton J, Arbuckle-Keil G (2013) Atomic force microscopy and micro-ATR-FT-IR imaging reveals fungal enzyme activity at the hyphal scale of resolution. Mycology 4:44–53

Pachler KGR, Matlok F, Gremlich H-U (1988) Merck FT-IR atlas. Wiley, Weinheim

Pallua JD, Recheis W, Pöder R, Pfaller K, Pezzei C, Hahn H, Huck-Pezzei V, Bittner LK, Schaefer G, Steiner E, Andre G, Hutwimmer S, Felber S, Pallua AK, Pallua AF, Bonn GK, Huck CW (2012) Morphological and tissue characterization of the medicinal fungus *Hericium coralloides* by a structural and molecular imaging platform. Analyst 137:1584–1595

Pandey KK, Pitman AJ (2003) FTIR studies of changes in wood chemistry following decay by brown-rot and white-rot fungi. Int Biodeterior Biodegradation 52:151–160

Peiris KHS, Bockus WW, Dowell FE (2012) Infrared spectral properties of germ, pericarp, and endosperm sections of sound wheat kernels and those damaged by *Fusarium graminearum*. Appl Spectrosc 66:1053–1060

Pena R, Lang C, Naumann A, Polle A (2014) Ectomycorrhizal identification in environmental samples of tree roots by Fourier-transform infrared (FTIR) spectroscopy. Front Plant Sci 229:1–9

Posch AE, Koch C, Helmel M, Marchetti-Deschmann M, Macfelda K, Lendl B, Allmaier G, Herwig C (2013) Combining light microscopy, dielectric spectroscopy, MALDI intact cell mass spectroscopy, FTIR spectromicroscopy and multivariate data mining for morphological and physiological bioprocess characterization of filamentous organisms. Fungl Genet Biol 51:1–11

Raab TK, Vogel JP (2004) Ecological and agricultural applications of synchrotron IR microscopy. Infrared Phys Technol 45:393–402

Rowlette J, Weida M, Bird B, Arnone D, Barre M, Day T (2014) High confidence, high-throughput screening with high-def IR microspectroscopy. BioOptics World 34–37. http://www.biopticsworld.com/articles/print/volume-7/issue-2/features/clinical-pathology-drug-development-live-cell-imaging-high-confidence-high-throughput-screening-with-high-def-ir-microspectroscopy.html

Salem EZ, Shahin IMI, Mohammed YF, Abdo HM, Abdel Hamid MF, Emam HE et al (2010) Applicability of Fourier transform infrared (FT-IR) spectroscopy for rapid identification of some yeasts and dermatophytes isolated from superficial fungal infections. J Egypt Women Dermatol Soc 7:105–110

Salman A, Pomerantz A, Tsror L, Lapidot I, Zwielly A, Moreh R, Mordechai S, Huleihel M (2011) Distinction of *Fusarium oxysporum* fungal isolates (strains) using FTIR-ATR spectroscopy and advanced statistical methods. Analyst 136:988–995

Salman A, Pomerantz A, Tsror L, Lapidot I, Moreh R, Mordechai S, Huleihel M (2012) Utilizing FTIR-ATR spectroscopy for classification and relative spectral similarity evaluation of different *Colletotrichum coccodes* isolates. Analyst 137:3558–3564

Salzer R, Siesler HW (eds) (2009) Infrared and Raman spectroscopic imaging, 1st edn. Wiley, Weinheim

Salzer R, Siesler HW (eds) (2014) Infrared and Raman spectroscopic imaging, 2nd edn. Wiley, Weinheim

Salzer R, Steiner G, Mantsch HH, Mansfield J, Lewis EN (2000) Infrared and Raman imaging of biological and biomimetic samples. Fresenius J Anal Chem 366:712–726

Santos C, Fraga ME, Kozakiewicz Z, Lima N (2010) Fourier transform infrared as a powerful technique for the identification and characterization of filamentous fungi and yeasts. Res Microbiol 161:168–175

Schulz H, Baranska M (2007) Identification and quantification of valuable plant substances by IR and Raman spectroscopy. Vib Spectrosc 43:13–25

Schulz H, Krähmer A, Naumann A, Gudi G (2014) Infrared and Raman spectroscopic mapping and imaging of plant materials. In: Salzer R, Siesler HW (eds) Infrared and Raman spectroscopic imaging. Wiley, Weinheim, pp 227–293

Schwarze FWMR (2007) Wood decay under the microscope. Fungal Biol Rev 21:133–170

Shapaval V, Schmitt J, Møretrø T, Suso HP, Skaar I, Åsli AW, Lillehaug D, Kohler A (2012) Characterization of food spoilage fungi by FTIR spectroscopy. J Appl Microbiol 114:788–796

Singh CB, Jayas DS, Borondics F, White NDG (2011) Synchrotron based infrared imaging study of compositional changes in stored wheat due to infection with *Aspergillus glaucus*. J Stored Prod Res 47:372–7

Szeghalmi A, Kaminskyj S, Gough KG (2007) A synchrotron FTIR microspectroscopy investigation of fungal hyphae grown under optimal and stressed conditions. Anal Bioanal Chem 387:1779–1789

Timmins ÉM, Howell SA, Alsberg BK, Noble WC, Goodacre R (1998) Rapid differentiation of closely related *Candida* species and strains by pyrolysis-mass spectrometry and Fourier transform-infrared spectroscopy. J Clin Microbiol 36:367–374

Toubas D, Essendoubi M, Adt I, Pinon J-M, Manfait M, Sockalingum GD (2007) FTIR spectroscopy in medical mycology: applications to the differentiation and typing of *Candida*. Anal Bioanal Chem 387:1729–1737

Vongsvivut J, Heraud P, Gupta A, Puri M, McNaughton D, Barrow CJ (2013) FTIR microspectroscopy for rapid screening and monitoring of polyunsaturated fatty acid production in commercially valuable marine yeasts and protists. Analyst 138:6016–6031

Weißhaupt P, Naumann A, Pritzkow W, Noll M (2013) Nitrogen uptake of *Hypholoma fasciculare* and coexisting bacteria. Mycol Prog 12:283–290

Wenning M, Seiler H, Scherer S (2002) Fourier-transform infrared microspectroscopy, a novel and rapid tool for identification of yeasts. Appl Environ Microbiol 68:4717–4721

Zervakis GI, Bekiaris G, Tarantilis PA, Pappas CS (2012) Rapid strain classification and taxa delimitation within the edible mushroom genus *Pleurotus* through the use of diffuse reflectance infrared Fourier transform (DRIFT) spectroscopy. Fungal Biol 116:715–728

Chapter 5
Whole Cells Imaged by Hard X-ray Transmission Microscopy

Zhiting Liang, Yong Guan, Shan Chen and Yangchao Tian

5.1 Introduction

The resolution of conventional light microscopes is diffraction limited. Super-resolution optical microscopies (Chap. 3) have been developed to reach tens of nanometer resolution but are only able to detect fluorescent labels (Huang 2010). Transmission electron microscopy (TEM) is used to analyze the structure of cells at sub-nanometer resolution (Chap. 8), but based on the low penetrating power of electrons, larger cells must first be sliced into 50–500-nm sections (Barcena and Koster 2009). Cryo focused ion beam (FIB) milling for serial block face imaging in the scanning electron microscope (SEM, Chap. 8) is an efficient method to achieve three-dimensional (3D) data of samples with a lateral pixel size of less than 7.5 nm and slice thickness better than 30 nm in Z (Schertel et al. 2013). With the development of high-brilliance synchrotron radiation light sources and high-precision X-ray-focusing optics, high spatial resolution transmission X-ray microscopy (TXM) is developing rapidly with significantly improved spatial resolution. Soft X-ray transmission microscopy (STXM) with a spatial resolution of about 12 nm has been reported (Chao et al. 2009), and hard X-ray microscopy resolution is down

Y. Tian (✉) · Z. Liang · Y. Guan · S. Chen
National Synchrotron Radiation Laboratory, University of Science and Technology of China, 3#317, No.42, Hezuohua South Road, Hefei, 230029 Anhui, P. R. China
e-mail: ychtian@ustc.edu.cn

Z. Liang
e-mail: zhiting@mail.ustc.edu.cn

Y. Guan
e-mail: yongg@ustc.edu.cn

S. Chen
e-mail: Chen3@mail.ustc.edu.cn

T. E. S. Dahms and K. J. Czymmek (eds.), *Advanced Microscopy in Mycology*,
Fungal Biology, DOI 10.1007/978-3-319-22437-4_5

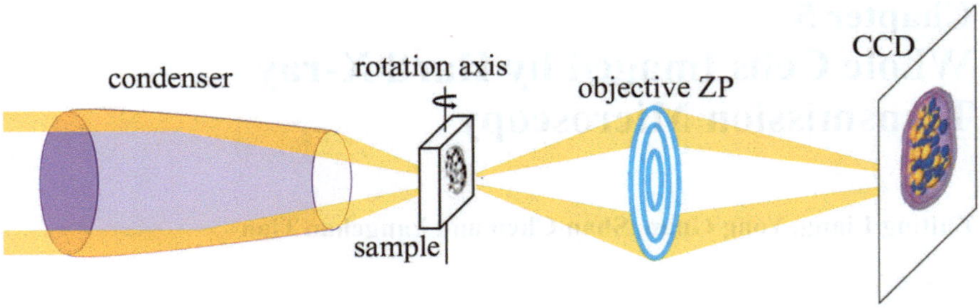

Fig. 5.1 Schematic of the soft X-ray transmission microscopy. Condenser: It focuses a hollow central cone of X-ray onto the sample through the inner surface reflecting. Rotation axis: Sample rotates around the axis to get projections at different angles. Objective zone plate *(ZP)*: It is used as an objective lens for X-ray microscopy. Charged-coupled device *(CCD):* It is a detector

to 30 nm (Chen et al. 2011). TXM has a number of advantages for the examination of biological specimens and material samples. One of the main advantages of this method is the ability to obtain a 3D image of the whole cell (rather than only fluorescently labeled structures), without the need for sectioning required by other methods (i.e., TEM).

TXMs can be divided into soft and hard X-ray microscopes depending on their working photon energy range. A schematic of the soft X-ray microscope is shown in Fig. 5.1. Soft X-ray microscopy was developed, in part, to provide a complementary spatial resolution between optical and electron microscopy and to make full use of the high natural absorption contrast between organic materials and water for photon energies between the K-absorption edge of oxygen ($E \approx 530$ eV or $\lambda = 2.34$ nm) and carbon ($E \approx 280$ eV or $\lambda = 4.43$ nm). The photon energies of the so-called "water window" are especially suitable for imaging samples in aqueous media. The only step required to prepare biological samples for cryo-STXM is plunging the sample into cryogenic liquid nitrogen for vitrification, to protect the sample from radiation damage (X-rays; Weiss et al. 2000) and vacuum inside the microscope chamber (Carzaniga 2013). Correlated cryo-STXM and cryo-fluorescence show that vitrifying the specimen significantly enhances the lifetime of the fluorophore (Le Gros 2009). Usually samples are screened by light microscopy, before transfer to the cryo-STXM sample station, to select appropriate cells, avoid cracks in the ice, debris, and thick ice (Duke 2013). Thus, hydrated biological specimens can be imaged in the photon energy range of the water window with no chemical fixation, dehydration, chemical staining, or physical sectioning. Cryo-STXM is used to visualize the internal architecture of fully hydrated cells (Scherfeld et al. 1998; Chichon et al. 2012; Muller et al. 2012; Hagen et al. 2012; Carrascosa et al. 2009; Drescher et al. 2013; Carzaniga et al. 2013; Hummel et al. 2012), including intact eukaryotic cells (Larabell and Le Gros 2004; Le Gros 2005; Parkinson et al. 2008; Parkinson et al. 2012; Uchida et al. 2009; McDermott et al. 2012) at high spatial resolution (50 nm or better). Recently, higher resolution STXMs have been developed and applied to reveal subcellular structures in whole cells (Schneider et al. 2010).

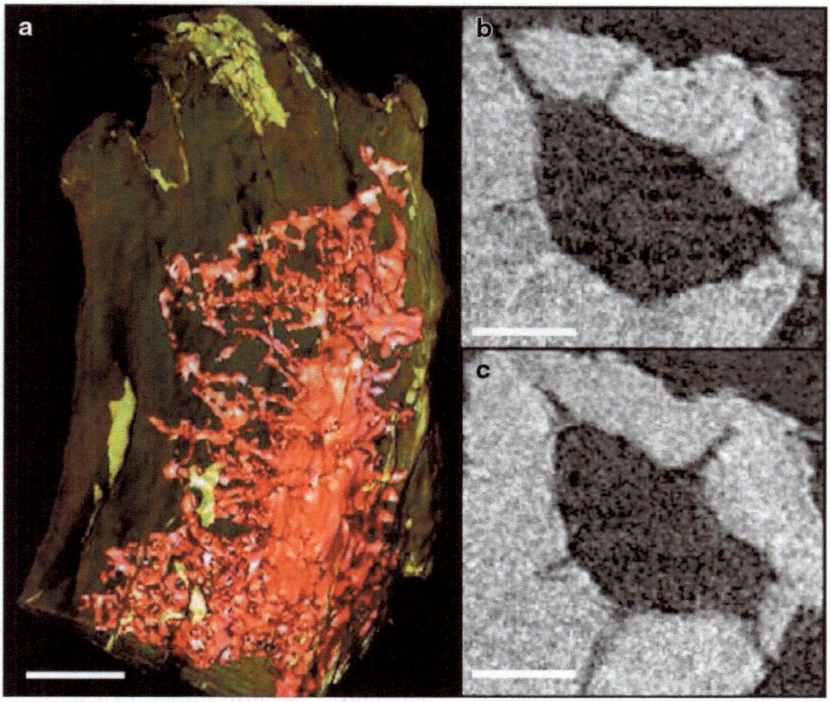

Fig. 5.2 **a** Three-dimensional reconstruction of a single lacuna surrounded by canalicular network (colored red) form mouse tibia trabecula (colored translucent yellow). **b**, **c** Reconstructed slices show processes extending from the lacuna. Scale bar is 5 μm in panel **a** and 2 μm in panels **b** and **c**. (Source: Andrews et al. 2010)

However, because of the low-penetration power of soft X-rays and the small depth of focus (DOF) of soft X-ray microscopy, biological sample size is limited, especially when 3D tomographic imaging is required. Thus, soft X-ray microscopy is not suitable for imaging large biological samples, including eukaryotic cells. Hard X-ray microscopy can overcome this limitation since it has high penetration and large DOF to facilitate 3D tomography of thicker samples with excellent resolution (Andrews et al. 2010). Hard X-ray transmission microscopy has been employed to observe various biological specimens (Yang et al. 2010; Zheng et al. 2012; Jeon et al. 2008; Zhang et al. 2013). Andrews et al. used hard X-ray transmission microscopy to image the internal structures and networks of mouse cancellous bone with high resolution (~30–40 nm; Andrews et al. 2010). Reconstructed data revealed the canalicular network including processes extending from the lacuna. Moreover, the 3D microstructure of a single lacuna found within a trabecula showed its shape, geometry, and surrounding canalicular network (Fig. 5.2a). The reconstructed slices (Fig. 5.2b, c) further show fine cell processes extending from lacunae in high-resolution detail (Andrews et al. 2010). In order to understand the internalization of

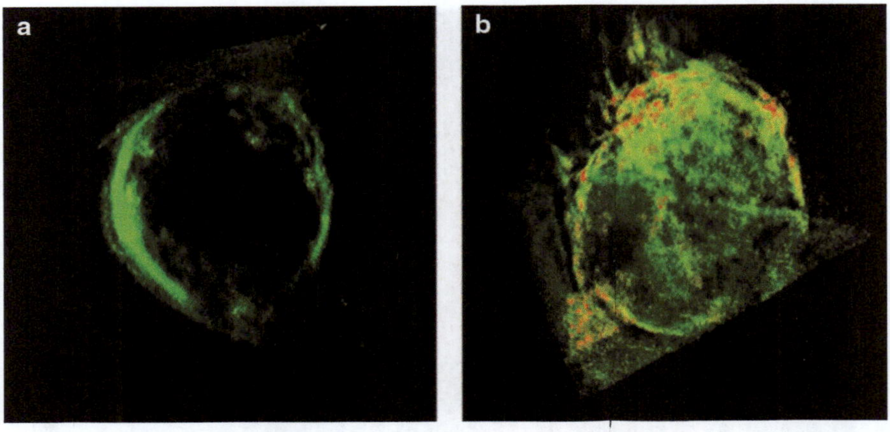

Fig. 5.3 Three-dimensional reconstructions of **a** a normal HeLa cell and **b** a cell after incubation with TiO_2 nanoparticles for 6 h (*red color* indicates the nanoparticle aggregates in the cell membrane). (Source: Zhang et al. 2013)

nanoparticles in cells, Cai et al. employed high-resolution hard X-ray transmission microscopy to investigate the cellular uptake and distribution of TiO_2 nanoparticles using HeLa cells. By rotating on the sample stage from −75 to +75° in 1° intervals, 151 projection images of a HeLa cell were collected. The 3D-reconstructed tomography maps of cells treated with TiO_2 nanoparticles showed that their aggregates were mainly distributed over the cell membrane surface (Fig. 5.3), which was confirmed by TEM. The results demonstrate that hard X-ray transmission microscopy has the ability to image the cellular distribution of nanoparticle aggregates and can be applied as a complementary analytical method to image the 3D distribution of nanoparticle aggregates in cells (Zhang et al. 2013).

Currently, hard X-ray tomography applied to biological imaging is in its initial stages, but results obtained to date demonstrate that it can provide a complementary tool for cellular imaging and analysis. In this chapter, we provide an overview of 3D microstructural analysis of yeast using hard X-ray transmission microscopy. Biological materials such as yeast have complex subcellular and intercellular structures, so their whole cell imaging requires a 3D technique on the nanometer scale with sufficient sample penetration. Hard X-ray microscopy meets these requirements by providing much better spatial resolution of cellular structures than light microscopy and by viewing internal structures of a whole cell without the need of sectioning. Moreover, when combined with tomographic procedures, hard X-ray microscopy can be used to make a complete 3D reconstruction of a whole cell to provide subcellular and intercellular microstructural information not available to conventional imaging techniques. Although the spatial resolution of hard X-ray microscopy is slightly less than that of soft X-ray microscopy, it can image large eukaryotic cells with thickness up to several tens of micrometers.

5.2 High-Resolution X-ray Microscopy

Since the refractive index of X-rays is close to unity for most materials, refractive lenses, similar to those used in light microscopy, are not applicable due to the strong absorption of X-rays by most materials. An ellipsoidal capillary has been developed as a condenser to focus X-rays, and a Fresnel zone plate is used as an objective lens. Figure 5.4 shows the layout of the transmission hard X-ray microscopy at the beamline U7A housed at the National Synchrotron Radiation Lab (NSRL) of China.

5.2.1 Light Source

When carrying out X-ray tomography, sample drift and noise caused by unavoidable instabilities in the microscope system will degrade the quality and clarity of the projected images and therefore the quality of final 3D tomographic reconstructions. Thus, it is important to collect the entire set of tomographic data with high signal-to-noise ratio in the shortest period of time, requiring a highly stable X-ray source such as that found at a synchrotron lab. When charged particles are accelerated in a radial fashion with bending magnets at the corners and insertion devices, wigglers, or undulators used in the straight sections of the storage ring (see Fig. 5.4), synchrotron radiation is produced. This radiation has a broad spectrum, which covers the infrared to hard X-ray regions and a high flux, which can allow rapid experiments. Thus, a synchrotron light source is the optimal source for high-resolution X-ray imaging. High-resolution X-ray microscopy facilities based on synchrotron radiation have been developed worldwide in the past decade. There are more than 47

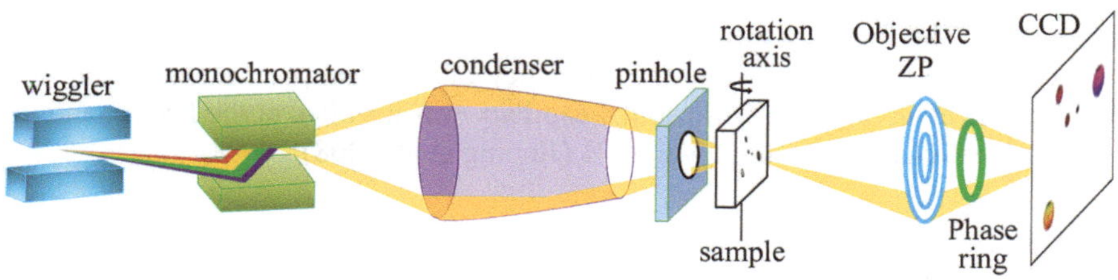

Fig. 5.4 The layout of the transmission X-ray microscopy *(TXM)* at the beamline U7A of the National Synchrotron Radiation Lab of China. Wiggler: It is composed of a series of magnets that are designed to periodically laterally deflect charged particles inside the storage ring of the synchrotron. These deflections create a change in acceleration, thus producing radiation with sufficient brightness and flux in the energy range of hard X-ray. Monochromator: It is an optical device that transmits a narrow band of wavelengths of light chosen from a wider range of wavelengths from the input. Pinhole: It is used as a monochromator to choose the wavelength range that can be transmitted. Phase ring: It causes a phase shift of the X-ray beam not diffracted by the sample, of either $\pi/2$ or $3\pi/2$. The phase contrast image is formed by the interference of the phase-shifted undiffracted light with the unshifted light diffracted by the sample light, translating phase modulations of the sample into intensity modulations in the image plane

synchrotron light sources worldwide (see http://lightsources.org/regions for their names, locations, and homepage). Most of the experimental end stations are open for applications from international users. Users can source appropriate equipment and their parameters from the homepage of each synchrotron, and then contact the person in charge of the end station to discuss related issues. The time available for doing an experiment depends on the operating conditions and the experimental arrangement at that moment. The NSRL storage ring operates with an electron beam energy of 0.8 GeV and a maximum current of 250 mA. A 6-T superconducting wiggler was installed at NSRL to produce hard X-rays with sufficient brightness and flux: 3.5×10^9 photons/s at a beam size of 12×13 mm^2 in the energy range of 7–12 keV. An existing beamline (U7A at NSRL), having a simple optical system designed to provide maximum output at a given energy resolution with acceptable cost, was selected to install the transmission X-ray microscope.

5.2.2 Focusing Optics

Ellipsoidal capillaries have been widely used as focusing optics for both soft and hard X-ray microscopy (Zeng et al. 2008; Guttmann et al. 2009). The ellipsoidal capillary condenser focuses a hollow central cone of illumination onto the sample and the objective zone plate. The source and focus are located at the two focal points of the ellipse generated when imaging the X-ray source with an ellipsoidal reflector. The desired parameters of the reflecting inner surface of the ellipsoidal capillary condenser are determined by two factors: the desired range of illumination angles on the output side and the distance between the X-ray source and the sample position (Zeng et al. 2008). The former should be designed to match the numerical aperture (NA) of the high-resolution objective zone plate. Glass capillaries with smooth inner surfaces can be drawn with a variety of fixed inner and outer diameters. The focusing efficiency of the capillary condenser is over 80 % due to lower surface roughness, while the efficiency of the zone plate is only near 10 % (Guttmann et al. 2009). An ellipsoidal capillary was installed at the NSRL, and the focusing efficiency measured was 85 % (Jinping et al. 2008).

5.2.3 X-ray Microscope Objective Lens

A Fresnel zone plate (Fig. 5.5), used as the objective lens in transmission X-ray microscopy, consists of a circular grating having a period that decreases radially in such a way that the diffraction orders from each zone add up coherently. The latter condition is achieved when all of the diffracted rays are in phase, that is, the optical path difference is $\pi/2$ between adjacent zones, opaque and transparent. The spatial resolution of an X-ray microscope depends on the numerical aperture (NA) of the objective zone plate given by:

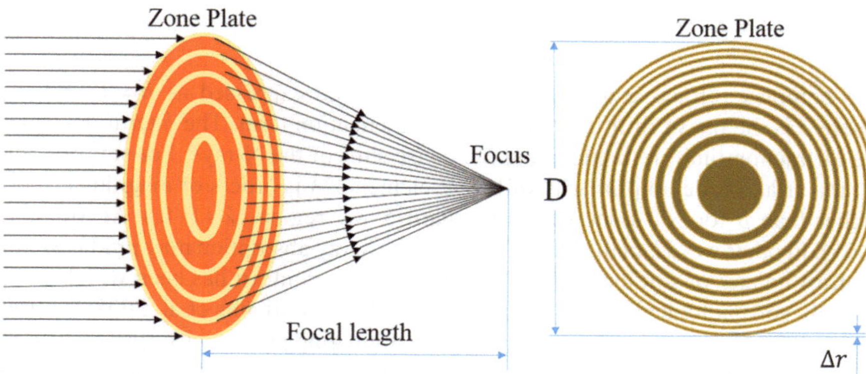

Fig. 5.5 Schematic drawing of a zone plate *(ZP)*. The diameter of the ZP is D and the width of the outermost zone is Δr. ZP is an objective lens for X-ray microscopy and Δr is the most crucial parameter of the system resolution

$$\delta = \frac{0.61}{NA} \times \frac{\lambda}{m} = \frac{1.22\Delta r}{m}, \tag{5.1}$$

where λ is the incident wavelength, m is the diffracted order, and Δr is the outermost zone width of the zone plate. According to the Rayleigh criterion, δ is the shortest distance between two features that can be resolved in the zone-plate image. The theoretical resolution of an X-ray microscope is proportional to the outermost zone width. The transmission X-ray microscope at NSRL has an objective zone plate lens with an 80 μm diameter and a 45-nm outer zone width, offering 55 nm resolution in the first-order diffraction for a photon energy of 8 keV (Hard X-rays wavelength, λ, of 0.15 nm). Nowadays, zone-doubled Fresnel zone plates with 20-nm outermost zone width are being designed and modeled to achieve a spatial resolution better than 20 nm at the Advance Photon Source at Argonne National Laboratory (Vila-Comamala et al. 2012).

Another objective zone plate parameter that needs to be considered is the DOF, defined by the distance over which the focus intensity is > 80 % of its maximum. The DOF is given by the following equation (Born and Wolf 1999):

$$\text{DOF} = \pm\lambda/\text{NA}^2 = \pm 2\Delta r^2/\lambda, \tag{5.2}$$

where NA is the numerical aperture of the zone plate; λ is the wavelength of the incident X-ray; and Δr is the outermost zone width of the objective zone plate. As demonstrated by this equation, when trying to achieve higher resolution, the DOF decreases. The transmission X-ray microscope at NSRL has a DOF of approximately 52 μm, enabling imaging of relatively thick samples.

5.2.4 The Zernike Phase Ring

The complex refractive index of the object can be expressed as $n = 1 - \delta + i\beta$, where δ is the phase shift term and β is the attenuation term. The δ term describes the diffraction limit, refraction, and the phase shift of the X-rays. For soft biological materials made up primarily of low-Z elements, δ is orders of magnitude (10^3) higher than β for soft materials in the hard X-ray region ($\lambda \approx 1-0.1$ Å; Momose 2005). In other words, in the hard X-ray region, X-ray absorption may produce poor contrast images of soft materials. Thus, X-ray phase contrast has been widely used to study soft biological tissues (Momose et al. 1996; Williams et al. 2008; Pfeiffer et al. 2007). Here, we use X-ray phase contrast imaging prior to absorption imaging of biological samples, which are weakly absorbing. In our recent imaging experiments, Zernike phase contrast was explored to visualize the 3D microstructure of yeasts (Yang et al. 2010; Zheng et al. 2012).

A phase ring (PR; see Figs. 5.4 and 5.6) is located at the back focal plane of the objective zone plate, causing a phase shift of the X-ray beam not diffracted by the sample, of either $\pi/2$ (positive phase contrast) or $3\pi/2$ (negative phase contrast). Positive phase contrast signifies greater optical thickness of the object, corresponding to a brighter image, while negative phase contrast has the opposite relationship (Kagoshima et al. 2001). The phase contrast image is formed by the interference of the phase-shifted undiffracted light with the unshifted light diffracted by the sample, translating phase modulations of the sample into intensity modulations in the image plane (see Fig. 5.6). At NSRL, the PR is made of Au with 2.4 μm thickness and 4 μm width (Tian et al. 2008), able to provide negative Zernike phase contrast. Figure 5.7 illustrates how the phase contrast image is improved over the absorption image, each taken at the same exposure time (Yang et al. 2010).

5.3 Tomographic Data Preprocessing and Reconstruction

5.3.1 Introduction to Tomography

There are many discretely and regularly distributed subcellular structures within a cell, for which their distribution and morphology relate to function. However, projection X-ray microscopy images of cells are two-dimensional (2D) in which many structural features are superimposed, making it difficult to observe and analyze cellular substructures. Therefore, acquiring high-resolution 3D images is desirable. When a series of projections are collected at angular intervals around a rotation axis, it is possible to reconstruct a 3D image of the sample using tomography.

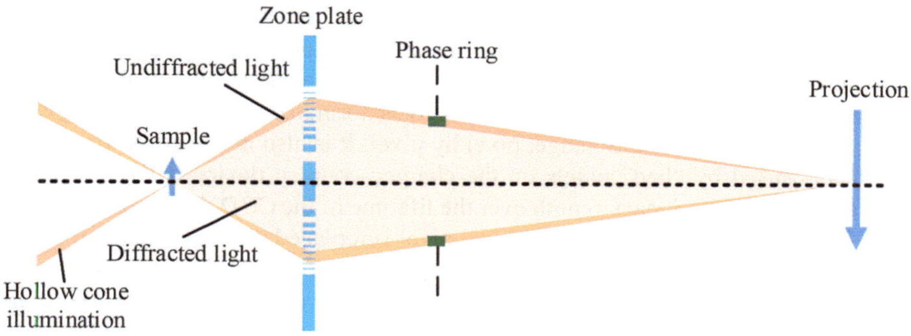

Fig. 5.6 Schematic of the annular phase ring. The phase ring is positioned following the ZP to intercept the undiffracted X-ray light. The thickness is designed to shift the phase of the undiffracted light by $3\pi/2$. The phase contrast image is formed by the interference of the phase-shifted undiffracted light with the unshifted light diffracted by the sample light, translating phase modulations of the sample into intensity modulations in the image plane

5.3.2 Tomographic Data Collection and Processing

For hard X-ray imaging, the projections are collected at 8 keV using the Zernike phase contrast model. When collecting projection images for tomographic reconstruction, the sample holder should permit the sample to remain within the field of view for isotropic data collection over 180° of rotation. However, a flat sample holder usually used for hard X-ray transmission microscopy can only be rotated ±74° before it occludes the X-rays, offering only 75 (or 150) projection images ranging from −74 to 74° in 2° (or 1°) intervals. The 3D internal microstructure of cells can completely be reconstructed based on these projections. While some image artifacts may arise from incomplete data collection, advanced reconstruction methods have been developed to attempt to eliminate such artifacts (Liang et al. 2013; Wang and Jiang 2004; Sidky et al. 2006).

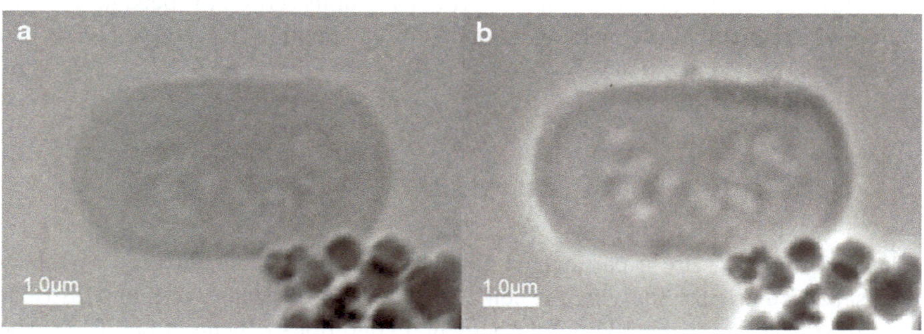

Fig. 5.7 Two models of yeast projections (Yang et al. 2010). **a** Absorption contrast projection of a yeast cell and **b** Zernike phase contrast projection of the same cell

Prior to reconstruction, the raw projections must first be processed. To correct for heterogeneities in sample illumination, which is due to the inhomogeneities of the light source, projections are normalized using a flat-field image collected with no sample in the field of view. Image intensity is normalized by dividing the projection image by the flat-field image, pixel by pixel. It is also necessary to correct for artifacts caused by "bad" pixels on the charged-coupled device (CCD), meaning those that have lost signal strength over the lifetime of the CCD. For each projection and flat-field image, the value of the bad pixel is replaced with a value interpolated from surrounding pixels. The instability of the rotation stage is accounted for by aligning all images to a common rotation axis prior to reconstruction. There are two main approaches: automated alignment (Parkinson et al. 2012) and manual alignment of the fiducial markers (i.e., 100-nm gold particles on the sample holder). Manual alignment is more accurate than automated, so the former is used to acquire initial results quickly, and higher-quality reconstructions can be achieved using the manual approach.

5.3.3 3D Reconstruction

Tomographic reconstruction is pursued once the images are properly processed. 3D sample images can be achieved by combining 2D projections. Numerous reconstruction algorithms have been proposed (Gordon et al. 1970; Andersen and Kak 1984; Raparia et al. 1998; Wang and Jiang 2004; Sidky et al. 2006; Andersen 1989; Zeng 2010), all with associated advantages and disadvantages. Generally speaking, reconstruction algorithms can be divided into two classes. The first class relies on Fourier methods and the central slice theorem (Zeng 2010), often the most popular choice based on the limited computational burden. The other class uses iterative reconstruction methods (Mueller et al. 1999; Liang et al. 2013; Cierniak 2011), which allow prior knowledge to be included in processing, and many regularizations (e.g., total variation minimization; Tian et al. 2011; Sidky and Pan 2008) and filters that can improve the quality of the results. Nevertheless, iterative reconstruction methods consume a large amount of computational time, preventing their widespread use. However, methods using graphics processing units (GPU) to speed up computation would enable these methods to regain more attention (Keck et al. 2009; Pan et al. 2009; Nett et al. 2010).

5.4 Cell Culture and Preparation

The fission yeasts are a particularly good eukaryotic cell model used for many years, particularly for studying the cell cycle and how the shape, size, and distribution of organelles change with environmental conditions (Zheng et al. 2012). The wild-type fission yeast, *Schizosaccharomyces pombe*, was cultured in liquid yeast extract peptone dextrose (YPD) medium at 25 °C for 12 h, rinsed in 0.2 M PIPES buffer (3 ×; piperazine-N, N'-bis[2-ethane sulfonic acid]) and fixed with 2% osmium tetroxide (in PIPES; 1 h, 25 °C), rinsed (2 ×) with PIPES buffer, stained with Reynolds' lead citrate (20 min), dehydrated in a graded ethanol series of 50–100% concentration (EtOH/(water + EtOH)), and then stained by 2% uranyl acetate in 95% ethanol solution (10 min). At last, the cells were spotted on a silicon nitride membrane of 100 nm thickness and air-dried. In this proof of principle, we chose the simplest drying method to show that hard X-ray tomography could be used to image cells. However, it is important to note that air-drying is not the best choice for cell sample preparation, causing drying artifacts, even when the sample had been fixed and ethanol substituted. In future, critical point drying (Lampert 1971), resin-embedded samples, and other new electron microscopy (EM) sample preparation technologies such as high-pressure freezing (McDonald 2007) and self-pressurized rapid freezing (Leunissen and Yi 2009), proven to maintain good sample shape, can be applied to hard X-ray microscopy.

5.5 Cell Organelles Identified

Projections were obtained using the phase contrast model and the photon energy of 8 keV. The 3D structure of the cell was reconstructed using a standard filtered back-projection algorithm (Kak et al. 1988) through a series of projections (Fig. 5.8a; Yang et al. 2010). Slice thickness is related to the pixel size of the CCD, approximately 58.3 nm for our system. The reconstructed cross-slices of cells from two different orientations are shown in Fig. 5.8b, c. The cell wall, with its three-layered structure, is clearly observed in the cross-slices (Fig. 5.8b, c, arrowheads), having a thickness of approximately 400 nm, which is typical for wild-type *S. pombe*. There are numerous small, near-spherical organelles having lower X-ray contrast than surrounding cytoplasm, and other organelles with irregular shape and higher X-ray contrast than the cytoplasm (Fig. 5.8b, c).

Segmentation can provide a way to isolate a specific cellular component from the data set, allowing 3D visualization of that component either alone or with other components. At present, identification of organelles by hard X-ray tomography is complex and difficult. Following heavy metal staining, every organelle of the cell will have a different X-ray absorption compared to the cytoplasm, allowing organelle boundaries to be identified. The density of lipid bodies and mitochondria is greater than the cytoplasm, so those components are more easily stained by heavy metal. Organelles having lower contrast (Fig. 5.8b, c) are consistent with vacuoles that also have low electron density in TEM studies (Konomi et al. 2003; Sajiki et al. 2009) that used comparable staining. Their morphology and distribution is also consistent with vacuoles identified by TEM and fluorescence microscopy (Bone et al. 1998; Mulvihill et al. 2001; Konomi et al. 2003; Takagi et al. 2003). The lower X-ray contrast organelles colored blue with near-spherical shape are identified as vacuoles, and the higher X-ray contrast organelles colored yellow are assigned as

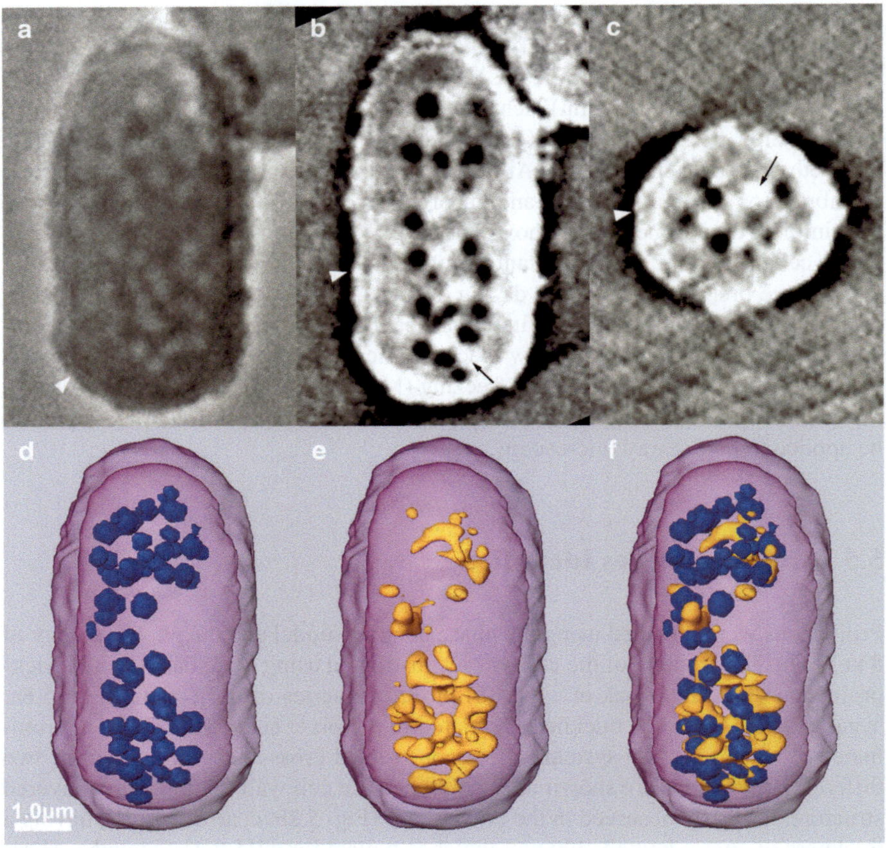

Fig. 5.8 Hard X-ray tomography of a fission yeast cell, *Schizosaccharomyces pombe* (Yang et al. 2010). **a** Projection of the yeast cell. **b**, **c** Reconstructing cross-slices through the cell from vertical orientations. *Arrows* point to the organelles that have higher X-ray contrast than surrounding cytoplasm, and *arrowheads* indicate the three-layer structure of the cell wall. **d**, **e** Organelles are divided into two groups based on their absorption contrast. The lower contrast organelles colored *blue* are identified as vacuoles, and the higher X-ray contrast organelles colored *yellow* are lipid bodies and mitochondria. **f** Total organization and distribution of organelles within a yeast cell

lipid bodies and mitochondria (Fig. 5.8). It is difficult to distinguish mitochondria and lipid bodies based on the hard X-ray tomographic data since resolution is limited, precluding exact organelle classification. By improving the resolution of hard X-ray microscopy in future, organelle identification will be more facile. In future work, the data could be correlated with TEM from the same sample to confirm the identity of mitochondria and lipid bodies based on their shape and size.

5.6 Quantitative Analysis of Organelles in the Cell

Cell size is a critical parameter in initiating cell division, and the number, morphology, and volume of the organelles have a profound influence on the function and viability of a cell (Uchida et al. 2011). Consequently, the quantitative analysis of cellular internal architecture is of great significance in cell science. Since hard X-ray tomography is capable of obtaining 3D architectures of biological samples, it has been used to investigate the yeast cycle at various stages. From the stationary to the growth phase, the shape of yeast transitioned from short and spherical (Fig. 5.9a) to long, thin, and elliptical (Fig. 5.9f). During cell division, a membrane (colored red) formed, and presumably septum cell wall components were also present, but their detection was beyond the resolution limit. The two cells had not yet separated completely (see Fig. 5.9m).

Yeast size is a decisive parameter for cell division, having to reach a minimum size before they begin to divide (Goranov et al. 2009; Jorgensen et al. 2002). During the cell cycle, organelle size must be precisely detected and controlled, both for proper cellular function and correct organelle partitioning (Neumann and Nurse 2007; Rafelski and Marshall 2008; Fagarasanu et al. 2007). Thus, in relation to cell growth and division, cell size, major organelle volume, and organelle to cell volume ratio were quantified (Table 5.1). As expected, cell volume gradually increases from the stationary phase (54.3 μm^3) to division (81.4 μm^3). The ratio of total major organelle to cell volume (O:C ratio) in each phase was 23.8, 10.6, and 11.1 % (Fig. 5.10), for stationary, growth, and division phases, respectively. A similar ratio was calculated for vacuoles, lipid bodies, and mitochondria alone (Fig. 5.10). From the growth to division phase, the O:C ratio for lipid bodies and mitochondria remained approximately stable (5.7–5.9 %) as did that for vacuoles (4.9–5.2 %). That is to say that the ratios existing between cell size and organelle volumes are consistent during the division and growth phases. On the contrary, the O:C ratio for all organelles, vacuoles, lipid bodies, and mitochondria in the stationary phase are higher than those during growth and division phases, indicating an increase in organelle volume during the stationary phase. Further, organelle shape varies in all three phases, with characteristic shapes in each phase. It can be concluded that hard X-ray microscopy is suitable for cellular imaging when some contrast enhancement method is introduced, such as Zernike phase contrast and heavy metal staining. Since samples were air-dried, shrinkage differences in the stationary and growth phases would influence the volume and ratios of the cells. Nonetheless, we demonstrated the ability of hard X-ray tomography to quantify and image cells. In future work, more sophisticated drying technologies developed for electron microscopy can be introduced to hard X-ray microscopy. The unique ability of hard X-ray microscopy to acquire images of whole, large cells in 3D could make it a complementary tool for imaging cells and providing useful information in cell biology (Yang et al. 2010).

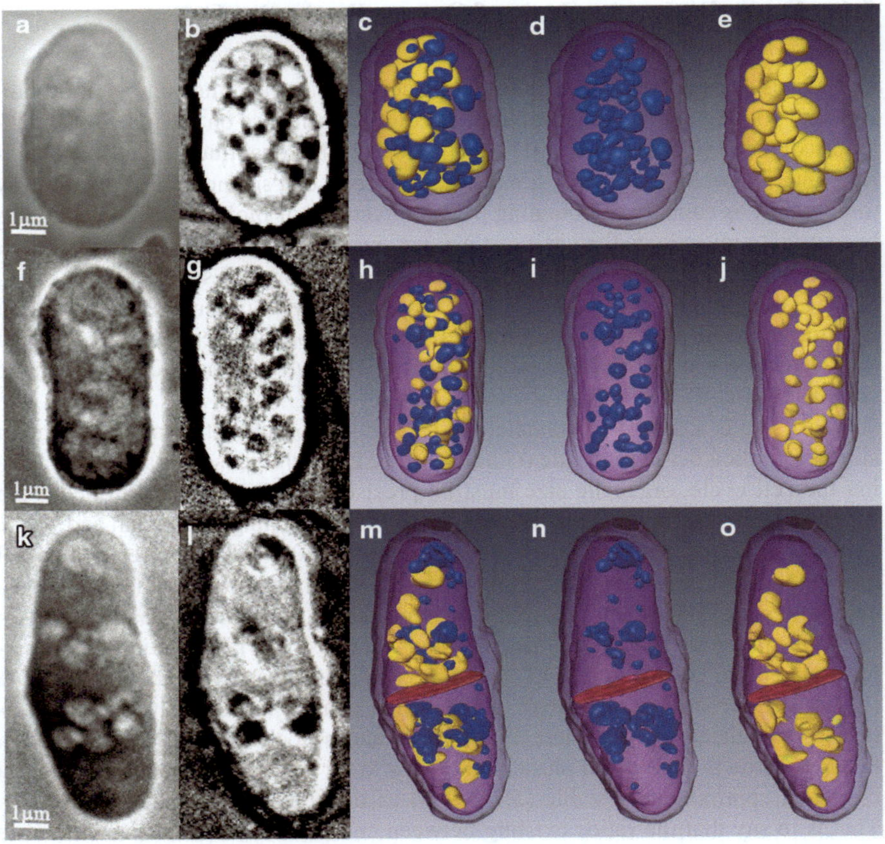

Fig. 5.9 Different phases of a fission yeast cell, *Schizosaccharomyces pombe,* imaged by hard X-ray tomography (Zheng et al. 2012). The three rows of images correspond to a cell in the stationary phase, a single adult cell, and a dividing yeast cell. The first column is the two-dimensional (2D) projection image, the second to fifth columns are reconstruction slices, the whole cells, vacuoles, and lipid bodies and mitochondria, respectively. (*m*, *n,* and *o*) A dense band between the mother and daughter cells is colored *red.* The organelles colored *blue* with near-spherical shape are identified as vacuoles, and the organelles colored *yellow* are assigned as lipid bodies and mitochondria

5.7 Future Developments of Hard X-ray Microscopy

In this chapter, we demonstrate that hard X-ray microscopy, using high penetration power and a large DOF, is a powerful tool for studying high-resolution 3D cell architecture. Contrast enhancement can be achieved with Zernike phase contrast and heavy metal staining. Based on the tomographic data, we can observe microstructural changes of yeasts during the cell cycle. While we demonstrate that hard X-ray microscopy and tomography have the ability to examine internal cellular ultrastructures, there is still room to improve spatial resolution. With the develop-

Table 5.1 Dimensions and volumes of organelles and cells from Fig. 5.9. (Source: Zheng et al. 2012)

Feature	Cell		
	Stationary	Growth	Division
Cell length (μm)	6.1	7.2	9.7
Cell width (μm)	4.1	3.5	5.1
Cell volume (μm^3)	54.3	68.2	81.4
Vol. organelles (μm^3)	12.9	7.2	10.0
Vol. lipid bodies and mitochondria (μm^3)	7.3	3.9	4.8
Vol. vacuoles (μm^3)	5.57	3.3	4.2

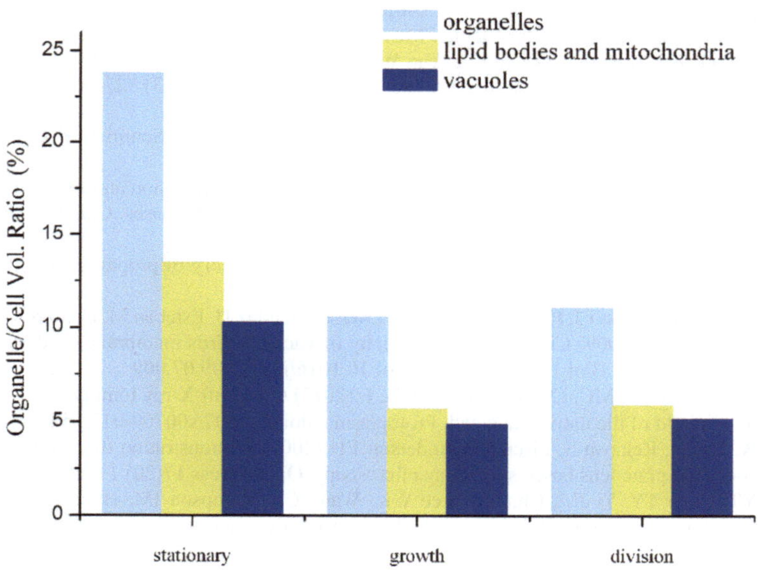

Fig. 5.10 The organelle to cell volume ratio, including major organelles, vacuoles, lipid bodies, and mitochondria, during **a** stationary phase, **b** cell growth, and **c** division phases of the cell cycle. (Source: Zheng et al. 2012)

ment of micro- and nanofabrication technology, the width of the outer zone of zone plates will be smaller, and the focusing diffraction efficiencies will be improved. In addition, the stability of the rotating station and data-processing software will also be promoted. Therefore, higher resolution will be achieved. Cryo-fixation, able to combat the radiation damage from X-rays and maintain the sample morphology, can be easily introduced into sample preparation methods for hard X-ray imaging. Furthermore, hard X-ray microscopy/tomography can be used in a correlative manner with other techniques to identify organelles. The data obtained by hard X-ray tomography can be correlated with electron microscopy of the same sample to identify organelles by morphology and position. Fluorescence imaging is another mature technology (Smith et al. 2013) that can be combined with hard X-ray microscopy to determine the position of fluorescently tagged molecules in the reconstruction images. Such integrative and correlative techniques will extend the potential of hard X-ray microscopy in cell imaging.

References

Andersen AH (1989) Algebraic reconstruction in CT from limited views. IEEE Trans Med Imaging 8(1):50–55. doi:10.1109/42.20361

Andersen A, Kak A (1984) Simultaneous algebraic reconstruction technique (SART): a superior implementation of the ART algorithm. Ultrason Imaging 6(1):81–94

Andrews JC, Almeida E, van der Meulen MC, Alwood JS, Lee C, Liu Y, Chen J, Meirer F, Feser M, Gelb J, Rudati J, Tkachuk A, Yun W, Pianetta P (2010) Nanoscale X-ray microscopic imaging of mammalian mineralized tissue. Microsc Microanal 16(3):327–336. doi:10.1017/S1431927610000231

Barcena M, Koster AJ (2009) Electron tomography in life science. Semin Cell Dev Biol 20 (8):920–930. doi:10.1016/j.semcdb.2009.07.008

Bone N, Millar JBA, Toda T, Armstrong J (1998) Regulated vacuole fusion and fission in *Schizosaccharomyces pombe*: an osmotic response dependent on MAP kinases. Curr Biol 8(3):135–144. doi:10.1016/S0960-9822(98)00060-8

Born M, Wolf E (1999) Principles of optics: electromagnetic theory of propagation, interference and diffraction of light. CUP Archive, New York

Carrascosa JL, Chichon FJ, Pereiro E, Rodriguez MJ, Fernandez JJ, Esteban M, Heim S, Guttmann P, Schneider G (2009) Cryo-X-ray tomography of vaccinia virus membranes and inner compartments. J Struct Biol 168(2):234–239. doi:10.1016/j.jsb.2009.07.009

Carzaniga R, Domart MC, Collinson LM, Duke E (2013) Cryo-soft X-ray tomography: a journey into the world of the native-state cell. Protoplasma. doi:10.1007/s00709-013-0583-y

Chao W, Kim J, Rekawa S, Fischer P, Anderson EH (2009) Demonstration of 12 nm resolution fresnel zone plate lens based soft X-ray microscopy. Opt Express 17(20):17669–17677

Chen YT, Chen TY, Yi JM, Chu YS, Lee WK, Wang CL, Kempson IM, Hwu Y, Gajdosik V, Margaritondo G (2011) Hard x-ray zernike microscopy reaches 30 nm resolution. Opt Lett 36(7):1269–1271

Chichon FJ, Rodriguez MJ, Pereiro E, Chiappi M, Perdiguero B, Guttmann P, Werner S, Rehbein S, Schneider G, Esteban M, Carrascosa JL (2012) Cryo X-ray nano-tomography of vaccinia virus infected cells. J Struct Biol 177(2):202–211. doi:10.1016/j.jsb.2011.12.001

Cierniak R (2011) Algebraic reconstruction techniques. In: X-ray computed tomography in biomedical engineering. Springer, London, pp 233–265

Drescher D, Guttmann P, Buchner T, Werner S, Laube G, Hornemann A, Tarek B, Schneider G, Kneipp J (2013) Specific biomolecule corona is associated with ring-shaped organization of silver nanoparticles in cells. Nanoscale 5(19):9193–9198. doi:10.1039/c3nr02129g

Duke EM, Razi M, Weston A, Guttmann P, Werner S, Henzler K, Schneider G, Tooze SA, Collinson LM (2013) Imaging endosomes and autophagosomes in whole mammalian cells using correlative cryo-fluorescence and cryo-soft X-ray microscopy (cryo-CLXM). Ultramicroscopy 143:77–87

Fagarasanu A, Fagarasanu M, Rachubinski RA (2007) Maintaining peroxisome populations: a story of division and inheritance. Annu Rev Cell Dev Biol 23:321–344. doi:10.1146/annurev.cellbio.23.090506.123456

Goranov AI, Cook M, Ricicova M, Ben-Ari G, Gonzalez C, Hansen C, Tyers M, Amon A (2009) The rate of cell growth is governed by cell cycle stage. Genes Dev 23(12):1408–1422. doi:10.1101/gad.1777309

Gordon R, Bender R, Herman GT (1970) Algebraic reconstruction techniques (ART) for three-dimensional electron microscopy and X-ray photography. J Theor Biol 29(3):471–481

Guttmann P, Zeng X, Feser M, Heim S, Yun W, Schneider G (2009) Ellipsoidal capillary as condenser for the BESSY full-field x-ray microscope. 9th international conference on X-Ray microscopy 186. doi:Artn 012064Doi 10.1088/1742-6596/186/1/012064

Hagen C, Guttmann P, Klupp B, Werner S, Rehbein S, Mettenleiter TC, Schneider G, Grunewald K (2012) Correlative VIS-fluorescence and soft X-ray cryo-microscopy/tomography of adherent cells. J Struct Biol 177(2):193–201. doi:10.1016/j.jsb.2011.12.012

Huang B (2010) Super-resolution optical microscopy: multiple choices. Curr Opin Chem Biol 14(1):10–14. doi:10.1016/j.cbpa.2009.10.013

Hummel E, Guttmann P, Werner S, Tarek B, Schneider G, Kunz M, Frangakis AS, Westermann B (2012) 3D ultrastructural organization of whole chlamydomonas reinhardtii cells studied by nanoscale soft X-ray tomography. PLoS ONE 7(12): e53293. doi:ARTN e53293DOI 10.1371/journal.pone.0053293

Jeon SY, Goo JW, Hong SP, Oh TH, Youn HS, Lee WS (2008) A new method for investigation of the hair shaft: hard X-ray microscopy with a 90-nm spatial resolution. Yonsei Med J 49(2):337–340. doi:10.3349/ymj.2008.49.2.337

Jinping T, Wenjie L, Jie C, Gang L, Ying X, Longhua L, Xinlong H, Yangchao T (2008) The design and test of ellipsoidal glass capillaries as condensers for X-ray microscope. Nucl Tech 31(09):671–675

Jorgensen P, Nishikawa JL, Breitkreutz BJ, Tyers M (2002) Systematic identification of pathways that couple cell growth and division in yeast. Science 297(5580):395–400. doi:10.1126/science.1070850

Kagoshima Y, Ibuki T, Yokoyama Y, Tsusaka Y, Matsui J, Takai K, Aino M (2001) 10 keV X-ray phase-contrast microscopy for observing transparent specimens. Jpn J Appl Phys 2 40 (11A):L1190–L1192. doi:10.1143/Jjap.40.L1190

Kak AC, Slaney M, IEEE Engineering in Medicine and Biology Society (1988) Principles of computerized tomographic imaging. IEEE Press, New York

Keck B, Hofmann H, Scherl H, Kowarschik M, Hornegger J (2009) GPU-accelerated SART reconstruction using the CUDA programming environment. In: SPIE medical imaging. International Society for Optics and Photonics, pp 72582B–72582B-72589

Konomi M, Fujimoto K, Toda T, Osumi M (2003) Characterization and behaviour of alpha-glucan synthase in *schizosaccharomyces pombe* as revealed by electron microscopy. Yeast 20(5):427–438. doi:10.1002/Yea.974

Lampert F (1971) Coiled supercoiled DNA in critical point dried and thin sectioned human chromosome fibres. Nat New Biol 234(49):187–188

Larabell CA, Le Gros MA (2004) X-ray tomography generates 3-D reconstructions of the yeast, *Saccharomyces cerevisiae*, at 60-nm resolution. Mol Biol Cell 15(3):957–962. doi:10.1091/mbc.E03-07-0522

Le Gros MA, McDermott G, Larabell CA (2005) X-ray tomography of whole cells. Curr Opin Struct Biol 15(5):593–600. doi:10.1016/j.sbi.2006.08.008

Le Gros M, McDermott G, Uchida M, Knoechel C, Larabell C (2009) High-aperture cryogenic light microscopy. J Microsc 235(1):1–8

Leunissen JLM, Yi H (2009) Self-pressurized rapid freezing (SPRF): a novel cryofixation method for specimen preparation in electron microscopy. J Microsc 235(1):25–35

Liang Z, Guan Y, Liu G, Bian R, Zhang X, Xiong Y, Tian Y (2013) Reconstruction of limited-angle and few-view nano-CT image via total variation iterative reconstruction. In: SPIE Optical Engineering + Applications, International Society for Optics and Photonics, pp 885113–885113–885117

McDermott G, Fox DM, Epperly L, Wetzler M, Barron AE, Gros MA L, Larabell CA (2012) Visualizing and quantifying cell phenotype using soft X-ray tomography. Bioessays 34(4):320–327. doi:10.1002/bies.201100125

McDonald K (2007) Cryopreparation methods for electron microscopy of selected model systems. Methods Cell Biol 79:23–56. doi:10.1016/S0091-679x(06)79002-1

Momose A (2005) Recent advances in X-ray phase imaging. Jpn J Appl Phys 44 (Part 1, 9 A):6355–6367. doi:10.1143/Jjap.44.6355

Momose A, Takeda T, Itai Y, Hirano K (1996) Phase-contrast X-ray computed tomography for observing biological soft tissues. Nat Med 2(4):473–475

Mueller K, Yagel R, Wheller JJ (1999) Anti-aliased three-dimensional cone-beam reconstruction of low-contrast objects with algebraic methods. IEEE Tran Med Imaging 18(6):519–537

Muller WG, Heymann JB, Nagashima K, Guttmann P, Werner S, Rehbein S, Schneider G, McNally JG (2012) Towards an atlas of mammalian cell ultrastructure by cryo soft X-ray tomography. J Struct Biol 177(2):179–192. doi:10.1016/j.jsb.2011.11.025

Mulvihill DP, Pollard PJ, Win TZ, Hyams JS (2001) Myosin V-mediated vacuole distribution and fusion in fission yeast. Curr Biol 11(14):1124–1127. doi:10.1016/S0960-9822(01)00322-0

Nett BE, Tang J, Chen GH (2010) GPU implementation of prior image constrained compressed sensing (PICCS). Medical imaging 2010: physics of medical imaging 7622. doi:10.1117/12.844578

Neumann FR, Nurse P (2007) Nuclear size control in fission yeast. J Cell Biol 179(4):593–600. doi:10.1083/jcb.200708054

Pan Y, Whitaker R, Cheryauka A, Ferguson D (2009) Feasibility of GPU-assisted iterative image reconstruction for mobile C-arm CT. In: SPIE medical imaging. International Society for Optics and Photonics, pp 72585J–72585J-72589

Parkinson DY, McDermott G, Etkin LD, Gros MA L, Larabell CA (2008) Quantitative 3-D imaging of eukaryotic cells using soft X-ray tomography. J Struct Biol 162(3):380–386. doi:10.1016/j.jsb.2008.02.003

Parkinson DY, Knoechel C, Yang C, Larabell CA, Le Gros MA (2012) Automatic alignment and reconstruction of images for soft X-ray tomography. J Struct Biol 177(2):259–266. doi:10.1016/j.jsb.2011.11.027

Pfeiffer F, Bunk O, David C, Bech M, Duc G L, Bravin A, Cloetens P (2007) High-resolution brain tumor visualization using three-dimensional X-ray phase contrast tomography. Phys Med Biol 52(23):6923–6930. doi:10.1088/0031-9155/52/23/010

Rafelski SM, Marshall WF (2008) Building the cell: design principles of cellular architecture. Nat Rev Mol Cell Biol 9 (8):593–602. doi:10.1038/nrm2460

Raparia D, Alessi J, Kponou A (1998) The algebraic reconstruction technique (ART). Proceedings of the 1997 Particle Accelerator Conference Vols 1–3:2023–2025

Sajiki K, Hatanaka M, Nakamura T, Takeda K, Shimanuki M, Yoshida T, Hanyu Y, Hayashi T, Nakaseko Y, Yanagida M (2009) Genetic control of cellular quiescence in S-pombe. J Cell Sci 122(9):1418–1429. doi:10.1242/Jcs.046466

Scherfeld D, Schneider G, Guttmann P, Osborn M (1998) Visualization of cytoskeletal elements in the transmission X-ray microscope. J Struct Biol 123(1):72–82. doi:10.1006/jsbi.1998.4017

Schertel A, Snaidero N, Han HM, Ruhwedel T, Laue M, Grabenbauer M, Mobius W (2013) Cryo FIB-SEM: volume imaging of cellular ultrastructure in native frozen specimens. J Struct Biol 184(2):355–360. doi:10.1016/j.jsb.2013.09.024

Schneider G, Guttmann P, Heim S, Rehbein S, Mueller F, Nagashima K, Heymann JB, Muller WG, McNally JG (2010) Three-dimensional cellular ultrastructure resolved by X-ray microscopy. Nat Methods 7(12):985–987. doi:10.1038/nmeth.1533

Sidky EY, Pan XC (2008) Image reconstruction in circular cone-beam computed tomography by constrained, total-variation minimization. Phys Med Biol 53(17):4777–4807. doi:10.1088/0031-9155/53/17/021

Sidky EY, C-MKa o, XcP n (2006) Accurate image reconstruction from few-views and limited-angle data in divergent-beam CT. J Xray Sci Technol

Smith EA, Cinquin BP, Do M, McDermott G, Gros MA L, Larabell CA (2013) Correlative cryogenic tomography of cells using light and soft X-rays. Ultramicroscopy 143:33–40. doi:10.1016/j.ultramic.2013.10.013

Takagi T, Ishijima SA, Ochi H, Osumi M (2003) Ultrastructure and behavior of actin cytoskeleton during cell wall formation in the fission yeast *schizosaccharomyces pombe*. J Electron Microsc 52(2):161–174. doi:10.1093/jmicro/52.2.161

Tian YC, Li WJ, Chen J, Liu LH, Liu G, Tkachuk A, Tian JP, Xiong Y, Gelb J, Hsu G, Yun WB (2008) High resolution hard x-ray microscope on a second generation synchrotron source. Rev Sci Instrum 79(10). doi:10.1063/1.3002484

Tian Z, Jia X, Yuan KH, Pan TS, Jiang SB (2011) Low-dose CT reconstruction via edge-preserving total variation regularization. Phys Med Biol 56(18):5949–5967. doi:10.1088/0031-9155/56/18/011

Uchida M, McDermott G, Wetzler M, Gros MA L, Myllys M, Knoechel C, Barron AE, Larabell CA (2009) Soft X-ray tomography of phenotypic switching and the cellular response to antifungal peptoids in Candida albicans. Proc Natl Acad Sci U S A 106(46):19375–19380. doi:10.1073/pnas.0906145106

Uchida M, Sun Y, McDermott G, Knoechel C, Gros MA L, Parkinson D, Drubin DG, Larabell CA (2011) Quantitative analysis of yeast internal architecture using soft X-ray tomography. Yeast 28(3):227–236. doi:10.1002/yea.1834

Vila-Comamala J, Pan Y, Lombardo JJ, Harris WM, Chiu WK, David C, Wang Y (2012) Zone-doubled fresnel zone plates for high-resolution hard X-ray full-field transmission microscopy. J Synchrotron Radiat 19 (5):705–709. doi:10.1107/S0909049512029640

Wang G, Jiang M (2004) Ordered-subset simultaneous algebraic reconstruction techniques (OS-SART). J Xray Sci Technol 12(3):169–177

Weiss D, Schneider G, Niemann B, Guttmann P, Rudolph D, Schmahl G (2000) Computed tomography of cryogenic biological specimens based on X-ray microscopic images. Ultramicroscopy 84(3):185–197

Williams IM, Siu KK, Gan R, He X, Hart SA, Styles CB, Lewis RA (2008) Towards the clinical application of X-ray phase contrast imaging. Eur J Radiol 68 (3 Suppl):73–77. doi:10.1016/j.ejrad.2008.04.042

Yang Y, Li W, Liu G, Zhang X, Chen J, Wu W, Guan Y, Xiong Y, Tian Y, Wu Z (2010) 3D visualization of subcellular structures of *schizosaccharomyces pombe* by hard X-ray tomography. J Microsc 240(1):14–20. doi:10.1111/j.1365-2818.2010.03379.x

Zeng GL (2010) Medical image reconstruction: a conceptual tutorial. Springer, New York

Zeng XH, Duewer F, Feser M, Huang C, Lyon A, Tkachuk A, Yun WB (2008) Ellipsoidal and parabolic glass capillaries as condensers for x-ray microscopes. Appl Opt 47(13):2376–2381. doi:10.1364/Ao.47.002376

Zhang JC, Cai XQ, Zhang Y, Li XM, Li WX, Tian YC, Li AG, Yu XH, Fan CH, Huang Q (2013) Imaging cellular uptake and intracellular distribution of TiO_2 nanoparticles. Anal Methods 5(23):6611–6616. doi:10.1039/C3ay41121d

Zheng T, Li WJ, Guan Y, Song XX, Xiong Y, Liu G, Tian YC (2012) Quantitative 3D imaging of yeast by hard X-ray tomography. Microsc Res Tech 75(5):662–666. doi:10.1002/Jemt.21108

Chapter 6
In Situ Nanocharacterization of Yeast Cells Using ESEM and FIB

Yajing Shen

Yeasts are eukaryotic microorganisms classified as fungi. As one of the earliest domesticated organisms, yeasts have been used for controlled fermentation of food and drinks and for leavening in baking throughout recorded history. Today, they are used in a variety of commercial fermentation and biomass conversion processes and in genetic engineering to produce certain hormones and enzymes important for wound healing and reducing inflammation.

6.1 *In Situ* Nanocharacterization System Inside the ESEM

The scanning electron microscope (SEM) is able to image samples at nanometer scale resolution and has been widely used for nanomaterial observations and characterization. Conventional SEM is fundamentally limited by the need for high vacuum conditions ($1.33 \times 10^{-3 \text{ to } -5}$ Pa), which prevents scattering of the primary electron beam throughout the system (Danilatos 1988). Thus, biological specimens are typically fixed and dried to maintain a structure that is as close as possible to its native state. Furthermore, SEM requires the specimen to be electrically conductive; otherwise a sample charge can quickly build up and impact/prevent the collection of high-resolution images. Therefore, to observe biological samples under an electron microscope, special sample processing has always been required.

In the 1970s, Danilatos and Robinson laid the ground work for the first SEM that did not require exceptionally low pressure, obviating the requirement of dried and coated specimens (Danilatos 1991). The first commercial environmental SEM

Y. Shen (✉)
Department of Mechanical and Biomedical Engineering, City University of Hong Kong, Tat Chee Avenue, Kowloon, Hong Kong, P. R. China
e-mail: yajishen@cityu.edu.hk

T. E. S. Dahms and K. J. Czymmek (eds.), *Advanced Microscopy in Mycology*, Fungal Biology, DOI 10.1007/978-3-319-22437-4_6

(ESEM) was produced by the late 1980s, offering a myriad of possibilities to image untreated, hydrated, and/or uncoated specimens. Philips FEI took over the commercialization of the instrument in 1996. Since then, the ESEM has been widely adopted by scientific, technical, and industrial researchers as evidenced by the exponential increase in publications arising from its use (Danilatos 1993).

ESEM enables imaging samples within a gas environment within the specimen chamber, setting ESEM apart from conventional SEM. The working concept of ESEM is based on two important principles: a differential pumping system and a gaseous scanning electron detector (GSED). Pressure-limiting apertures separate the electron microscope (EM) column into discrete pressure zones so that the electron gun and upper parts of the column can be held at high vacuum (1.33×10^{-5} Pa), while the pressure within the specimen chamber can be maintained up to 2.67 kPa. The GSED is used for general wet imaging and for high-pressure imaging with auxiliary gases. ESEM allows imaging in the presence of water vapor within the sample chamber, allowing maintenance of a high relative humidity so that samples can be imaged in a hydrated state without the need for dehydration and fixation. Therefore, ESEM provides a novel approach for imaging biological samples, including wood fibers, leaf tissue, and cell surfaces with labeled proteins (Muscariello et al. 2008; Shaler and Mott 1996; Zheng et al. 2008). Compared with traditional optical microscopes, the ESEM is not diffraction limited, offering resolution at the nanometer scale. Compared to the atomic force microscope (AFM, see Chap. 7), ESEM has a larger field of view, does not require direct sample contact, and facilitates real-time observation.

A nanomanipulation system provides an effective solution for *in situ* biological sample characterization inside the ESEM chamber. Rather than only sample imaging, the nanomanipulation system enables ESEM as an analytical tool. For example, a glass probe manipulated precisely to compress the yeast cell has been used to measure its stiffness (Ren et al. 2008). We constructed a flexible nanomanipulation system inside the ESEM (Fig. 6.1), which consists of three units: Unit 1 and 2 are for nanomanipulation, and unit 3 is the cooling stage. Unit 1 has three degrees of freedom (DOF) with linear translational motions in the *x*-, *y*-, and *z*-axes, while unit 2 has four—three linear nanopositioners in the *x*-, *y*-, and *z*-axes and one rotational positioner about the *z*-axis. The sample stage mounted on the cooling stage allows for temperature control (~0 to ~40 °C).

Biological samples can be studied under wet conditions with a relatively high chamber pressure (10–2600 Pa) in ESEM mode. For sample preparation, yeast cell suspensions are first deposited onto the cooling stage. Then, by gradually decreasing the pressure in the ESEM chamber water is evaporated, allowing imaging of yeast cells. As shown in Fig. 6.2, virtually all yeast cells remained spherical due to the high relative humidity (up to 100 %).

The focused ion beam (FIB) technique (see also Chap. 8) was largely developed during the 1970s through the early 1980s, with the first commercial instrument appearing in 1990 (Reyntjens and Puers 2001). Now, it is a powerful nanofabrication method using both top-down and bottom-up approaches. There are three basic

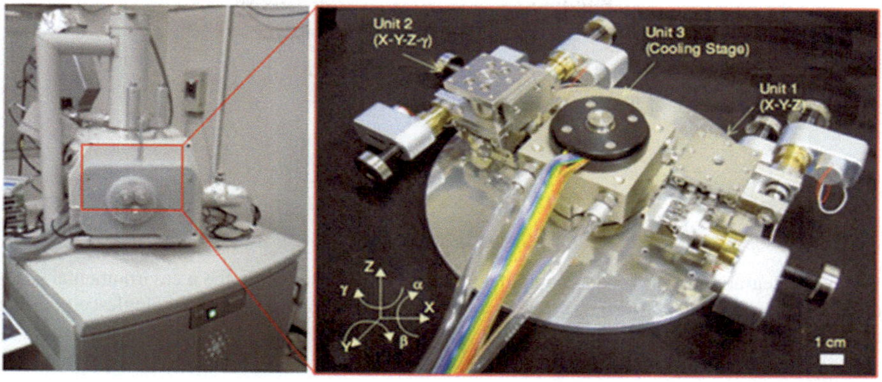

Fig. 6.1 Nanomanipulation system inside the environmental scanning electron microscope (ESEM). (Reprinted with permission from Shen et al. 2011a)

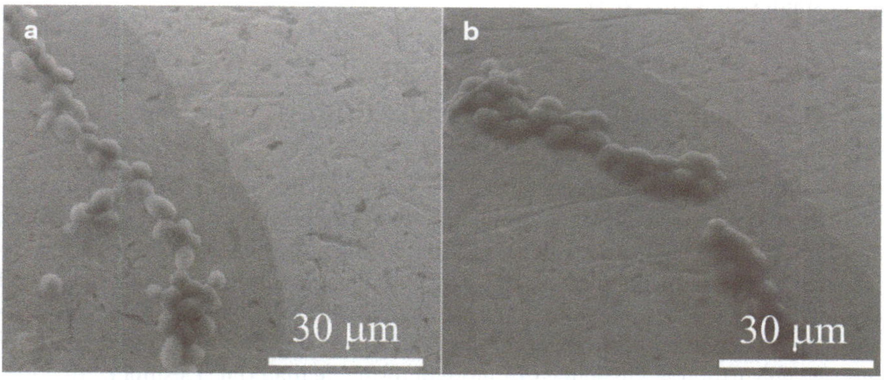

Fig. 6.2 Scanning electron microscope (SEM) images of yeast cells in environmental SEM (ESEM) mode. Acceleration voltage = 15 kV. **a** Chamber temperature *(T)* = 0.5 °C and pressure *(P)* = 444 Pa for 70 % humidity. **b** Sample chamber T=0.5 °C and P=630 Pa for 98 % humidity

operations of the FIB: (1) enhancing observation of the specimen surface by working in scanning ion microscopy mode, (2) etching the specimen surface with the ion beam, and (3) depositing gaseous material on the specimen surface using the ion beam. In the following sections, the nanomanipulation tools fabricated by FIB are applied to yeast cell characterization.

In summary, ESEM, with an image resolution up to 3.5 nm, is an excellent way to observe surface characteristics of biological samples. The nanomanipulator offers a linear resolution of 30 nm to drive and control precision movement and access

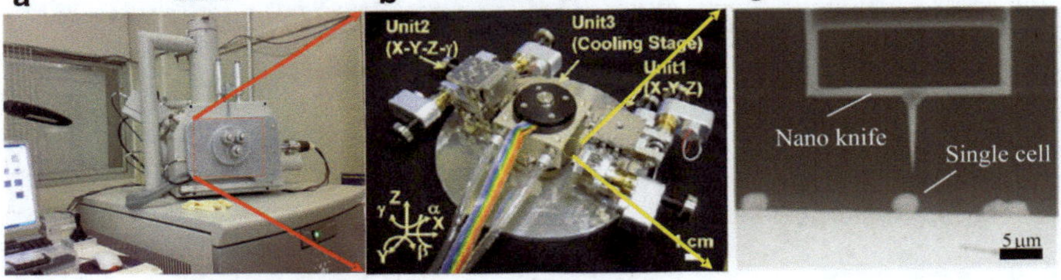

Fig. 6.3 Integration of the **b** nanomanipulator and a **c** nanoknife within the **a** environmental scanning electron microscope *(ESEM)* as an *in situ* nanocharacterization system for single-cell cutting and manipulation

single cells. The functional nanotools allow measurement and characterization of the cell's properties. Therefore, the integration of ESEM with the nanorobotic manipulation system and nanotools provides a powerful approach for yeast cell analysis at small scales (Fig. 6.3).

6.2 Characterization of Yeast Cell Stiffness

Cell stiffness is closely related to the structure of the cell wall and the cytoskeleton, a protein-based cellular scaffolding in the cytoplasm (Ning et al. 1993; Wang and Sun 2012). Thus, cell stiffness reflects cellular function in many aspects, for instance, stiffness of the mammalian cell will change with drug treatment or cell death (Li et al. 2011; Lam et al. 2007; Sarah et al. 2011). Change in plant cell stiffness can reflect cuticular cracking in ripe fruit, an area of fruit skin having integrity failure and surrounded by a mass of more deformable material (Desmet et al. 2003; Matas et al. 2004). In this section, a method is presented to measure the stiffness of individual yeast cells based on the integrated ESEM system. A nanoneedle sensor, fabricated by FIB etching, was housed on a buffering beam and a nanomanipulation system used to drive deformation of individual cells measured by ESEM. A mathematical model was developed to describe cell stiffness, which was calculated from experimental data (see Sect. 1.2.2).

6.2.1 Nanoneedle Fabrication by FIB

A schematic drawing of the nanoneedle used to measure cell stiffness is shown in Fig. 6.4. The nanoneedle, fabricated from a commercial atomic force microscopy (AFM) cantilever by FIB etching (Fig. 6.4A), is rounded at the tip with a diameter of approximately 200 nm used to indent the cell surface under the force of the nanomanipulator within the ESEM. A narrow buffering beam was designed that houses

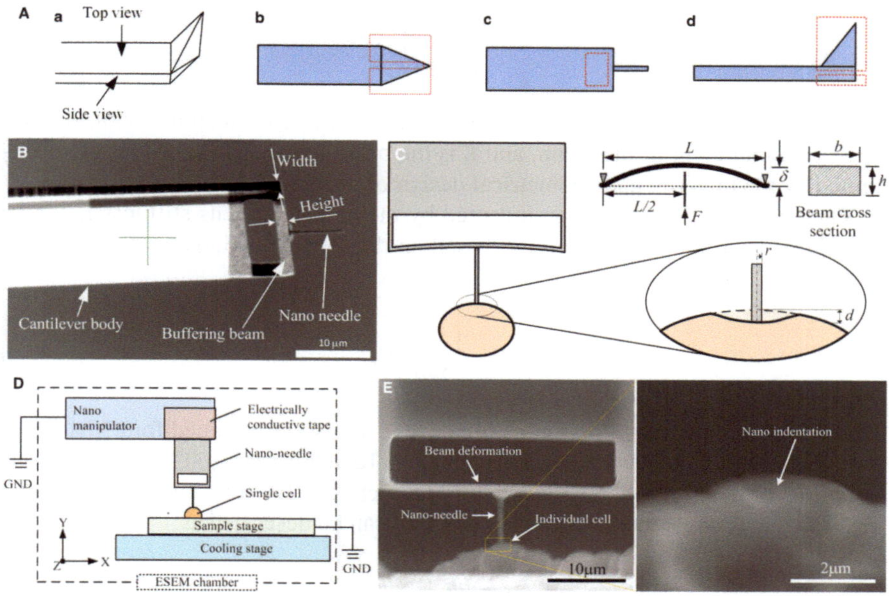

Fig. 6.4 **A** Focused ion beam (FIB) was used to fabricate the nanoneedle from a commercial atomic force microscope (AFM) cantilever (OMCL-AC240TM, Olympus Inc.). Shown in **a** is a schematic of the AFM cantilever, **b** etching to fabricate the nanoneedle, **c** formation of the narrow buffering beam by etching a rectangular area near the needle, and **d** rotation of the cantilever 90° for side etching. **B** Scanning electron microscope (SEM) images of the nanoneedle. **C** Schematic representation of the mathematical model representing the nanoneedle indentation of a single cell. **D** Schematic of the yeast cell stiffness measurement inside the environmental SEM *(ESEM)* by deformation with the beam-mounted nanoneedle. **E** Beam and cell deformation during the measurement process. *Left image:* Deformation of the nanoneedle beam and single cell. *Right image:* Magnified image of the cell indentation. Pressure: 261 Pa, temperature: 0.5 °C, humidity: 40 %. *GND* ground. (Reprinted with permission from Shen et al. 2013)

the nanoneedle and senses the applied force. The length of the buffering beam is approximately 24 µm, the height is 0.4 µm, and the width is 2.8 µm (Fig. 6.4B). Compared to a commercial AFM cantilever, the length ratio of the nanoneedle is much smaller, and the buffering beam design prevents tip twisting and damage during touchdown.

6.2.2 *Mathematical Model to Determine Stiffness*

The nanoneedle can be simply modeled as a beam with a load on the center, as shown in Fig. 6.4C (Shen et al. 2013). Therefore, the relationship between the loading force F and the deformation of the loading point δ can be described based on the Euler–Bernoulli beam theory:

$$F = \frac{4E_b bh^3}{L^3}\delta, \tag{6.1}$$

where E_b is the elastic modulus of the beam, b is the cross section width, h is the cross section height of the beam, and L is the beam length. These values are determined by the material and geometrical design of the nanoneedle.

Once the yeast cell has been indented by the nanoneedle, its stiffness E can be calculated from the loading force F based on the Hertz–Sneddon model (Sneddon 1965):

$$E = \frac{1-\upsilon^2}{2rd}F, \tag{6.2}$$

where υ, r, and d are the Poisson's ratio ($\upsilon = 0.5$ is always assumed for a yeast cell; Svaldo Lanero et al. 2006; Stenson et al. 2009; Touhami et al. 2003), the radius of a cylindrical tip, and the cell's indentation, respectively. By substituting Eq. 6.1 into Eq. 6.2, the Young's modulus of the yeast cell can be described as:

$$E = \frac{2E_b bh^3(1-\upsilon^2)}{rL^3}\cdot\frac{\delta}{d}, \tag{6.3}$$

where E_b, b, h, υ, r, and L are constant values and, therefore, δ and d are the only two parameters that need to be measured.

6.2.3 Proof of Principle: Cell Stiffness Measurement

A schematic drawing of the single-cell stiffness measurement experiment is shown in Fig. 6.4D. Yeast cells were placed on the sample stage, and the nanoneedle was housed on the nanomanipulator within the ESEM chamber. During the measurement, the nanoneedle is moved to the top of a yeast cell and then pressed against the cell wall. Cell deformation is determined during the elastic deformation period when the beam mount of the nanoneedle is bent. These deformations were observed and measured directly from the ESEM image, as shown in Fig. 6.4E. As a result, the stiffness of the yeast cell can be calculated based on Eq. 6.2. In this demonstration, the beam is deformed 250 nm, and the cell is indented 267 nm, for which the stiffness of the cell is calculated to be 15.5 MPa.

Cell stiffness is regulated by its biological condition, such as the stage of growth or development and its relative health. The mechanical property of a cell can be affected by its environment, which includes temperature, humidity, and other parameters. This characterization approach, using a nanoneedle inside the ESEM, provides a novel way to investigate the mechanical properties of the cell with high resolution. Although the ESEM environment does not represent a native state for the yeast cell, it offers mechanical characterization of yeast as a material, which could potentially benefit basic yeast cell research and biotechnological applications. More details regarding these cell stiffness measurements can be found elsewhere (Shen et al. 2013).

6.3 Cell–Substrate Adhesion Strength

The adhesion between cells and substrates, or extracellular matrices, plays an important role in biological processes. For example, there can be a strong correlation between adhesion strength and cell morphology, growth, differentiation, and function (Marcotte and Tabrizian 2008). Furthermore, cell–substrate adhesion studies are critical for the development of biocompatible materials in biomedical engineering (Razatos et al. 1998; Dubas et al. 2009). Here, we present a method for measuring adhesion strength with our modified ESEM nanocharacterization system. A "microputter" was fabricated from an AFM cantilever by FIB, a mathematical model for the adhesion force measurement was established, and the experiment was demonstrated inside the ESEM. One application of the cell–substrate adhesion measurement was to investigate cell viability, as one factor in the quantified adhesive force.

6.3.1 Preparation of the Microputter

The AFM cantilever is a powerful tool for cell–substrate adhesion studies owing to its small size and force-sensing capability. However, the traditional AFM cantilever has a sharp tip, approximately 15 nm, which can damage the cell when using it to detach the cell from the substrate directly. Therefore, as illustrated in Fig. 6.5A, a microputter with a flat tip (2×2 μm rectangular area) was fabricated from the AFM cantilever using FIB etching. The spring constant of the microputter could be controlled by modifying the structure of the cantilever beam as shown in Fig. 6.5.

The spring constant of the microputter was calibrated using a nanomanipulation approach. As the images show in Fig. 6.5B, the microputter presses a reference cantilever driven by the nanomanipulator. Deflection of the cantilever and the microputter beam was measured from ESEM images, allowing calculation of the microputter spring constant. Further detail of the calibration process can be found elsewhere (Shen et al. 2011c). To demonstrate the method, two microputters with different spring constants (2.07 ± 0.16 and 0.15 ± 0.02 N/m) were prepared and calibrated.

6.3.2 Mathematic Model for Adhesion Force Measurements

As the SEM images show in Fig. 6.5C, the microputter was attached to the nanomanipulator, and yeast cells were placed on the substrate and then pushed and detached by the microputter. During the detachment process, the beam of the microputter has

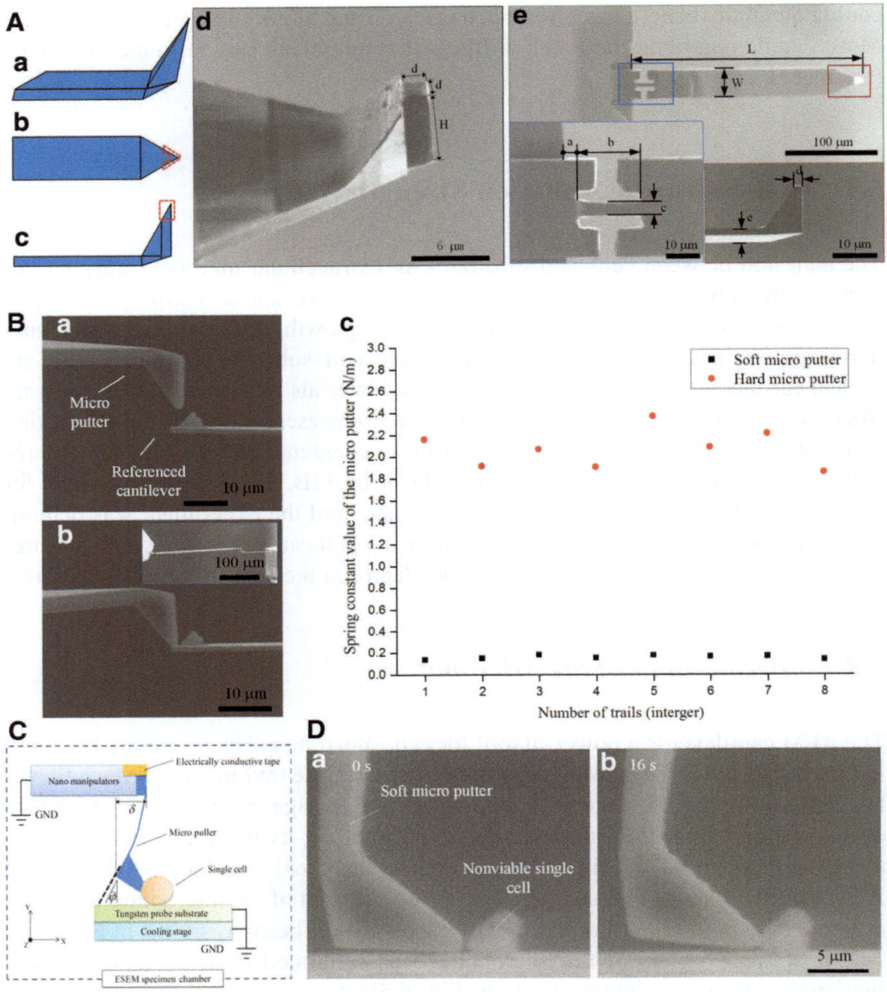

Fig. 6.5 Preparation of the microputter by focused ion beam (FIB). **A** Hard and soft microputter. a=4.0 μm, b=20.0 μm, c=4.5 μm, d=2.0 μm, e=2.8 μm, L=240 μm, W=30 μm. **B** Spring constant calibration by nanomanipulation. **C** Schematic of the cell–substrate adhesion force measurement. **D** Scanning electron microscope (SEM) images of the single nonviable cell adhesion force measurement using the soft microputter **a** before and **b** during detachment. *ESEM* environmental SEM, *GND* ground. (Reprinted with permission from Shen et al. 2011c)

a deflection δ, which can be measured from the ESEM images directly. Thus, the value of adhesion force F can be calculated by Hooke's law directly:

$$F = k\delta, \tag{6.4}$$

where k is the spring constant of the microputter, and δ is the deflection of the tip.

6.3.3 Cell Viability as a Function of Adhesive Force

The adhesion of a cell to a matrix relates to cellular activity and biological function. Since cell adhesion molecules are not expressed by nonviable cells, the adhesive force between a single nonviable cell and a substrate is expected to be reduced compared to viable cells. Therefore, adhesion strength may represent a parameter identifying cell viability, which is very important for studies involving for example toxicology, drug screening, and food quality inspection.

Viable and nonviable yeast cell colonies were confirmed by vital staining (Kucsera et al. 2000), and the adhesion force between substrates and the viable and nonviable yeast colonies were measured using the microputter. As ESEM images show in Fig. 6.5D, the microputter significantly deflects during the cell pushing, representing the adhesion strength between the nonviable cell and the tungsten substrate. To ensure the reproducibility, three sterilized substrates (i.e., tungsten, ITO (indium tin oxide), and gold) were used, having adhesion forces with single viable cells of 17.4 ± 4.7, 21.2 ± 4.4, and 19.4 ± 3.4 μN, respectively. However, the respective adhesion forces of single nonviable cells on the same three substrates were only 6.1 ± 1.8, 7.2 ± 2.4, and 3.4 ± 1.7 μN, statistically ($p = 0.05$, $n = 7$) smaller, approximately threefold, than for viable cells.

These results provide evidence that the interaction between the cell and extracellular matrices can be a useful quantitative parameter to evaluate the biological condition of cells, a novel way to study yeast cell mechanics at the nanoscale.

6.4 Cell–Cell Adhesion Strength

The adhesion strength between individual cells is related to the cell's biological activities (Gumbiner 1996). Studies on cell–cell adhesion would benefit a more in-depth understanding of biological processes. In this section, a method is introduced for studying cell–cell adhesion using the ESEM nanocharacterization system.

6.4.1 Schematics of the Experimental Setup

Figure 6.6 illustrates the experimental setup for cell–cell adhesion force measurements using ESEM nanocharacterization. Here, two nanorobotic manipulators were employed, named NM1 and NM2 on which the nanopicker and microprobe were housed, respectively (Fig. 6.6a). During the measurement process, the microprobe was first used to transfer one cell on top of the target cell (Fig. 6.6b–e). Next, the nanopicker tip was inserted into the cell–cell interface and was used to lift the top cell until separation (Fig. 6.6e–h). The adhesion force F was calculated based on the bounce height of the nanopicker using Hooke's law.

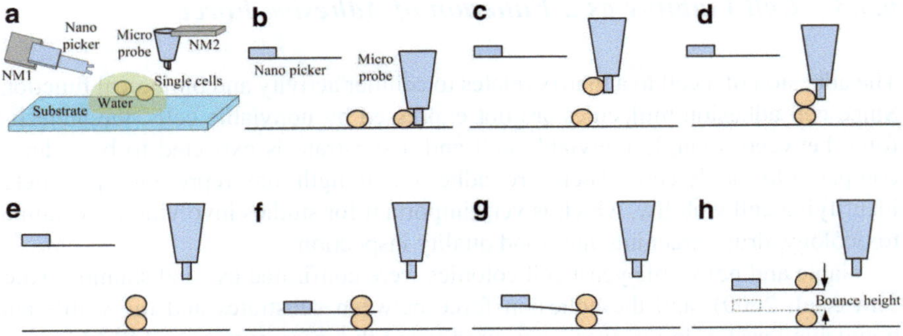

Fig. 6.6 Schematic drawing of the cell–cell adhesion force measurement using the nanopicker inside the environmental scanning electron microscope (ESEM). **a** Water solution containing yeast cells is pipetted on the substrate surface. The nanopicker with a V-shaped tip is housed on nanomanipulator 1 *(NM1),* and the microprobe is housed on nanomanipulator 2 *(NM2).* **b–e** A single cell is transferred to the top of another cell by the microprobe. **f–h** The adhesion force is measured by the nanopicker based on the bounce height δ. In the experiment, the relative humidity around the cell is set to 70 % by setting the chamber pressure and temperature to 444 Pa and 0.5 °C, respectively. (Reprinted with permission from Shen et al. 2011b)

6.4.2 Cell–Cell Adhesive Force Measurements

As the images show in Fig. 6.7A, the V-shaped nanopicker was fabricated from a commercial AFM cantilever using FIB etching. The open side and end of the V-shaped tip was approximately 4 and 2 μm, respectively, allowing the tip to be inserted into the cell–cell interface to lift the cell until they completely separate (Fig. 6.7B). The measurement results show that the adhesive force of the cell/nanopicker is much smaller than that between cells and so can be ignored (Fig. 6.7C). The results indicate that adhesive force initially relates to contact time (i.e., it sharply increases to 25 nN during the first 5 min), but then subsequently stabilizes. The ESEM nano-characterization system provides an effective approach to study cell–cell adhesion strength at the nanoscale. Additional details on the fabrication of the nanopicker and the measurement method can be found elsewhere (Shen et al. 2011b).

6.5 Yeast Cell Cutting: Towards Next Generation of Nanosurgery

The ability to slice cells into small pieces is crucial in cell biology research, such as for high-resolution observation of inner cell structure (Hess 2007; Leis et al. 2009; Möbius 2009). Moreover, as a method to manipulate the cell, single-cell slicing is essential for next-generation nanosurgery but comes with the challenge of limiting cell damage. This section reviews a nanoknife with a buffering beam designed for *in situ* single-cell slicing. Three nanoknives, with various edge angles, were fabricated

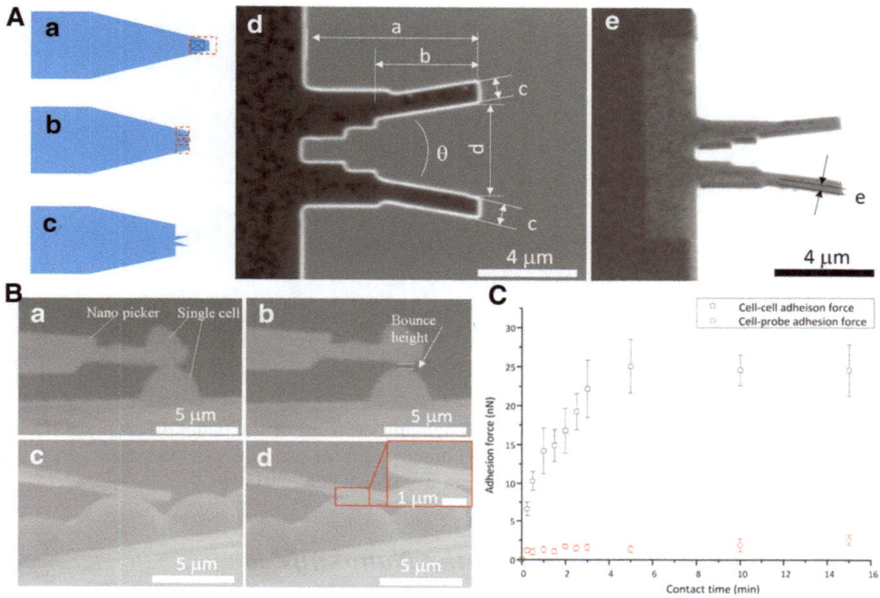

Fig. 6.7 Demonstration of cell–cell adhesion force measurement. **A** Preparation of the nanopicker: $a=7.0$ μm, $b=4.2$ μm, $c=0.7$ μm, $d=4.1$ μm, $e=0.4$ μm. $\theta=20°$. **B** Experiment with **a** and **b** the cell–cell adhesion force measurement, and **c** and **d** evaluation of the cell–probe adhesion. **a** and **c** show the initial contact conditions, while **b** and **d** show the bounce height after separation. **C** Adhesive forces between cell and probe and between cells as a function of contact time. (Reprinted with permission from Shen et al. 2011b)

from AFM cantilevers by FIB for *in situ* single yeast cell slicing with the nanocharacterization system inside the ESEM. Cutting forces and single-cell slice angle after cutting were evaluated.

6.5.1 Preparation of the Nanoknife

A schematic drawing of the nanoknife design is shown in Fig. 6.8A. This kind of nanoknife can provide a very sharp tip compared with the traditional diamond or glass knife and is expected to result in less cell compression during cutting. A buffering beam fabricated to support the nanoknife protects it from damage in the event that the knife tip touches the substrate unexpectedly. The nanoknife fabrication process was similar to that of the nanoneedle presented in the previous section. To investigate the effect of the knife angle on cell cutting quality, three nanoknives with different angles (5, 25, and 45°) were fabricated by the FIB etching process shown in Fig. 6.8B.

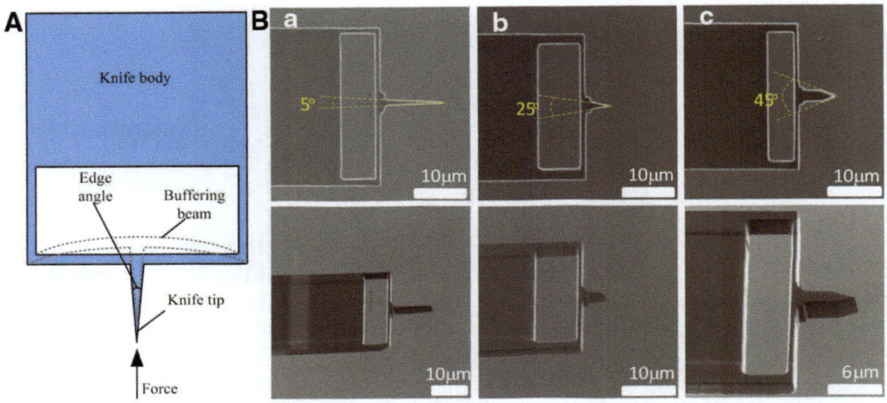

Fig. 6.8 Preparation of the nanoknives. **A** Schematic drawing of the nanoknife with buffering beam. **B** Scanning electron microscope (SEM) images of three nanoknives with various angles: **a** 5°, **b** 25°, and **c** 45°. (Reprinted with permission from Shen et al. 2011d)

6.5.2 *Single-Cell Slicing with the Nanoknife*

The *in situ* single-cell slicing experiment was performed using the three nanoknives (Fig. 6.9) in which the cutting force and the slice angle were used to evaluate slicing efficiency. Cutting force was calculated based on the mathematical model described in Eq. 6.1, and the slice angle was measured from the ESEM directly.

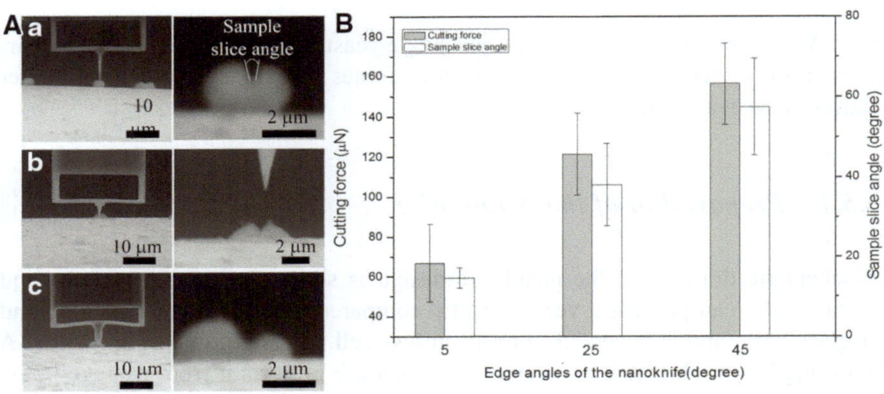

Fig. 6.9 Slicing of single cells using a nanoknife within the environmental scanning electron microscope (ESEM). **A** Cell cutting using the nanoknife with different angles: **a** 5, **b** 25, and **c** 45°. **B** experimental results. The *gray bar* and the *left* *y*-axis label indicate cutting force, while the *white bar* and the *right* *y*-axis label indicate sample slice angle. (Reprinted with permission from Shen et al. 2011d)

The cutting force and sample slice angle of each knife are shown in Fig. 6.9. As expected, the cutting force and the sample slice angle were smaller with the knife having the smaller edge angle. For instance, the average cutting force (66.7 μN) was over 50% smaller for the 5° nanoknife compared to the 45° knife (156.3 μN). The average sample slice angle was approximately 15° using the 5° nanoknife and increased to 57° with the 45° knife.

The nanoknife with a sharp angle provides a better solution for cell slicing. It reduces the compression to the cell, minimizing mechanical damage of the slice structure, potentially providing more precise information on cell structure. Moreover, cell slicing with the nanoknife is direct and does not require sample prepreparation, such as freezing and embedding. Therefore, it could be used to track mechanical and physiological responses to cell wounding *in situ*. Last but not least, the nanoknife provides an ideal nanotool for less invasive single-cell surgery, which may have significant application in future. A more detailed description of the nanoknife can be found elsewhere (Shen et al. 2011d).

6.6 Conclusions and Future Prospects

This chapter presents yeast cell *in situ* characterization by ESEM and FIB, which combines nanoobservation, nanomanipulation, and micro-/nanofunctional tools. This combination of instruments provided real-time images at high resolution while manipulating the yeast cell precisely in three dimensions. Many functional micro-/nanotools can be fabricated by FIB to measure yeast cell stiffness, cell–substrate adhesion, cell–cell interactions, and cell slicing by ESEM with potential applications to many other fields. Moreover, these nanotools can be applied to other instruments, such as the ambient pressure environment of high-resolution optical microscopy or with fluorescent dyes for fluorescence microscopy. Such flexibility would greatly promote its application in biological fields.

As will be introduced in Chap. 7, AFM is a powerful tool for fungal studies. The AFM can image the cell surface at nanometer resolution and measure the mechanical properties at each point on the piconewton scale, creating a map of stiffness/elastic distribution of the cell or cell wall. However, most AFM systems lack the ability to manipulate the cell during characterization, and it is difficult to replace the AFM cantilever with other nanotools. Since some of the nanotools described herein are fabricated from AFM cantilevers, they may be applicable to AFM, allowing flexible manipulation, such as cell detachment and cell cutting.

ESEM/FIB offers novel opportunities for fungal studies, especially those that require highly accurate *in situ* observation and manipulation and could significantly complement optical microscopy for current characterization techniques.

There are still many opportunities for ESEM/FIB to be applied in future, including nanoscale control strategies, sensor fusion for biological analysis, object identification, and evaluation, to name a few. A high-efficiency and intelligent characterization system can greatly improve experimental efficiency, extricating scientists from the complex manipulation process. In addition to mechanical characterization, other properties, such as pH, temperature, and chemistry, could be probed provided the appropriate nanotools are developed. Since the micro-/nanotools fabricated by FIB etching usually have a single characterization function, they are not suitable for multipurpose characterization, making the fabrication of multifunctional nanotools critical for fully investigating the cell. One possible solution is to combine traditional microelectromechanical systems (MEMS) with FIB fabrication (i.e., roughly fabricating the main body by MEMS and modifying the end effector precisely by FIB). Information obtained by the ESEM system could be integrated with that from optical and atomic force microscopy. Although ESEM is a powerful system for nanocharacterization, it lacks a great deal of information that could be supplemented by other microscopic methods. Combining information from different systems (see Chap. 8) can prove powerful for understanding biological processes. For example, an optical microscope integrated into the SEM chamber has produced some preliminary fluorescent images and high-resolution surface images of *Caenorhabditis elegans* (Nakajima et al. 2011). Further advances are sure to materialize in the near future.

References

Danilatos G (1988) Foundations of environmental scanning electron microscopy. Adv Electron Electron Phys 71:109–250

Danilatos G (1991) Review and outline of environmental SEM at present. J Microsc 162:391–402

Danilatos G (1993) Introduction to the ESEM instrument. Microsc Res Tech 25:354–361

Desmet M, Lammertyn J, Scheerlinck N, Verlinden BE, Nicolaï BM (2003) Determination of puncture injury susceptibility of tomatoes. Postharvest Biol Technol 27:293–303

Dubas ST, Kittitheeranun P, Rangkupan R, Sanchavanakit N, Potiyaraj P (2009) Coating of polyelectrolyte multilayer thin films on nanofibrous scaffolds to improve cell adhesion. J Appl Polymer Sci 114:1574–1579

Gumbiner BM (1996) Cell adhesion: the molecular basis of tissue architecture and morphogenesis. Cell 84:345–357

Hess M (2007) Cryopreparation methodology for plant cell biology. Methods Cell Biol 79:57–100

Kucsera J, Yarita K, Takeo K (2000) Simple detection method for distinguishing dead and living yeast colonies. J Microbiol Methods 41:19–21

Lam WA, Rosenbluth MJ, Fletcher DA (2007) Chemotherapy exposure increases leukemia cell stiffness. Blood 109:3505–3508

Leis A, Rockel B, Andrees L, Baumeister W (2009) Visualizing cells at the nanoscale. Trends Biochem Sci 34:60–70

Li M, Liu L, Xi N, Wang Y, Dong Z, Tabata O et al (2011) Imaging and measuring the rituximab-induced changes of mechanical properties in B-lymphoma cells using atomic force microscopy. Biochem Biophys Res Commun 404:689–694

Marcotte L, Tabrizian M (2008) Sensing surfaces: challenges in studying the cell adhesion process and the cell adhesion forces on biomaterials. IRBM 29:77–88

Matas AJ, Cobb ED, Bartsch JA, Paolillo DJ, Niklas KJ (2004) Biomechanics and anatomy of *Lycopersicon esculentum* fruit peels and enzyme-treated samples. Am J Bot 91:352–360

Möbius W (2009) Cryopreparation of biological specimens for immunoelectron microscopy. Ann Anat (Anatomischer Anzeiger) 191:231–247

Muscariello L, Rosso F, Marino G, Barbarisi M, Cafiero G, Barbarisi A (2008) Cell surface protein detection with immunogold labelling in ESEM: optimisation of the method and semi-quantitative analysis. J Cell Physiol 214:769–776

Nakajima M, Hirano T, Kojima M, Hisamoto N, Homma M, Fukuda T (2011) Direct nano-injection method by nanoprobe insertion based on E-SEM nanorobotic manipulation under hybrid microscope. In: Robotics and Automation (ICRA), 2011 IEEE International Conference on, pp 4139–4144

Ning W, James PB, Donald EI (1993) Mechanotransduction across the cell surface and through the cytoskeleton. Science 260:21

Razatos A, Ong YL, Sharma MM, Georgiou G (1998) Molecular determinants of bacterial adhesion monitored by atomic force microscopy. Proc Natl Acad Sci U S A 95:11059

Ren Y, Donald AM, Zhang Z (2008) Investigation of the morphology, viability and mechanical properties of yeast cells in environmental SEM. Scanning 30:435–442

Reyntjens S, Puers R (2001) A review of focused ion beam applications in microsystem technology. J Micromech Microeng 11:287

Sarah EC, Yu-Sheng J, Qing-Yi L, JianYu R, James KG (2011) Green tea extract selectively targets nanomechanics of live metastatic cancer cells. Nanotechnology 22:215101

Shaler S, Mott L (1996) Microscopic analysis of wood fibers using ESEM and confocal microscopy. In: Proceeding of the woodfiber plastic composites, vol 25, p 32

Shen Y, Ahmad MR, Nakajima M, Kojima S, Homma M, Fukuda T (2011a) Evaluation of the single yeast cell's adhesion to ITO substrates with various surface energies via ESEM nanorobotic manipulation system. IEEE Trans Nanobioscience 10:217–224

Shen Y, Nakajima M, Kojima S, Homma M, Fukuda T (2011b) Study of the time effect on the strength of cell–cell adhesion force by a novel nano-picker. Biochem Biophys Res Commun 409:160–165

Shen Y, Nakajima M, Kojima S, Homma M, Kojima M, Fukuda T (2011c) Single cell adhesion force measurement for cell viability identification using an AFM cantilever-based micro putter. Meas Sci Technol 22:115802

Shen Y, Nakajima M, Yang Z, Kojima S, Homma M, Fukuda T (2011d) Design and characterization of nanoknife with buffering beam for in situ single-cell cutting. Nanotechnology 22:305701

Shen Y, Nakajima M, Yang Z, Tajima H, Najdovski Z, Homma M et al (2013) Single cell stiffness measurement at various humidity conditions by nanomanipulation of a nano-needle. Nanotechnology 24:145703

Sneddon IN (1965) The relation between load and penetration in the axisymmetric Boussinesq problem for a punch of arbitrary profile. Int J Eng Sci 3:47–57

Stenson J, Thomas C, Hartley P (2009) Modelling the mechanical properties of yeast cells. Chem Eng Sci 64:1892–1903

Svaldo Lanero T, Cavalleri O, Krol S, Rolandi R, Gliozzi A (2006) Mechanical properties of single living cells encapsulated in polyelectrolyte matrixes. J Biotechnol 124:723–731

Touhami A, Nysten B, Dufrêne YF (2003) Nanoscale mapping of the elasticity of microbial cells by atomic force microscopy. Langmuir 19:4539–4543

Wang K, Sun D (2012) Influence of semiflexible structural features of actin cytoskeleton on cell stiffness based on actin microstructural modeling. J Biomech 45:1900–1908

Zheng Y, Han D, Zhai J, Jiang L (2008) In situ investigation on dynamic suspending of microdroplet on lotus leaf and gradient of wettable micro-and nanostructure from water condensation. Appl Phys Lett 92:084106-084106-3

Chapter 7
Imaging Living Yeast Cells and Quantifying Their Biophysical Properties by Atomic Force Microscopy

Cécile Formosa and Etienne Dague

7.1 Introduction

Yeasts, such as *Saccharomyces cerevisiae,* were used for thousands of years by humans to produce food and beverages, but humans have also cohabited with harmful yeasts such as *Candida albicans*. Both are surrounded by a thick, mechanically strong cell wall that plays several key physiological roles, such as maintaining cell shape and cell integrity, protecting the cell interior from harmful xenobiotics in the environment. The cell wall also harbors several proteins that are implicated in molecular recognition and adhesion (Chaffin 2008). The chemical composition of the yeast cell wall is well known (Lipke and Ovalle 1998). It consists of a microfibrillar network of β-glucans (β-1,3 and β-1,6-glucans) that represent 50–60 % of the cell wall mass, overlaid by highly glycosylated proteins that are decorated by long chains of mannose residues representing 40–50 % of the cell wall mass. Chitin, a linear polysaccharide of β-linked N-acetylglucosamine, is another major component of the yeast cell wall and represents 1–3 % of the cell wall mass. The yeast cell wall is an essential organelle for cell viability because it preserves the cell from osmotic pressure and heat shock, and it serves as a barrier and a filter against harmful molecules. Interestingly, among eukaryotes, this cell wall is unique to fungi, and, as

E. Dague (✉) · C. Formosa
CNRS, LAAS, 7 avenue du Colonel Roche, 31400 Toulouse, France
e-mail: edague@laas.fr

Université de Toulouse; LAAS, 31400 Toulouse, France

C. Formosa
CNRS, UMR 7565, SRSMC, Vandœuvre-lès-Nancy, France

Faculté de Pharmacie, Université de Lorraine, UMR 7565, Nancy, France

T. E. S. Dahms and K. J. Czymmek (eds.), *Advanced Microscopy in Mycology*,
Fungal Biology, DOI 10.1007/978-3-319-22437-4_7

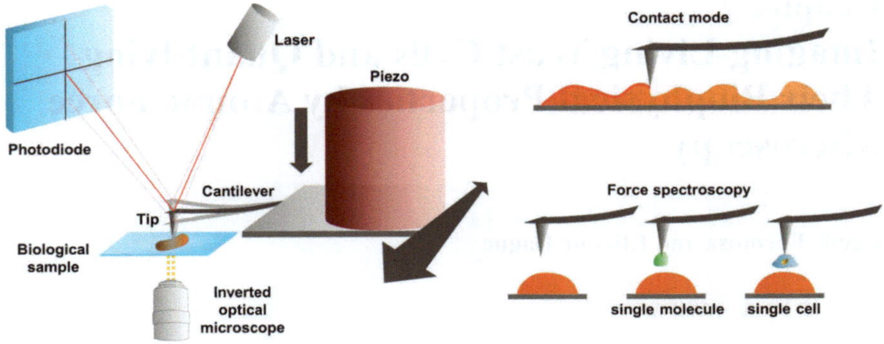

Fig. 7.1 A sharp tip is mounted on a cantilever that can be moved in the *x, y,* and *z* direction by a piezoelectric ceramic. The deflection of the cantilever is monitored on a four-quadrant photodiode as the reflection of a laser beam aligned at the end of the, usually gold-coated, cantilever. The AFM can be used to produce topographical images (i.e., contact and tapping modes with raster scanning) or to measure forces (force spectroscopy mode) between a bare or a functionalized (biomolecule or single-cell) tip and the sample. (Reprinted with permission from Pillet et al. 2014a)

it is essential for yeast viability, it represents an excellent target for antifungal drugs targeted against pathogenic yeasts. In addition, the molecular architecture of the yeast cell wall is not static, but constantly remodeled as a function of growth conditions, morphological development, and in response to cell surface stresses (Orlean 2012). Therefore, atomic force microscopy (AFM), which visualizes and probes the ultrastructure of the cell wall surface, is perfectly suited for the study of its dynamic structure and its molecular modification under different conditions.

Since its invention in 1986 (Binnig and Quate 1986), AFM has been used more and more to explore living cells at the nanoscale. AFM provides the unique opportunity to measure topography, nanomechanical properties, and/or single-molecule interactions, on living cells, with a nanoscale resolution. In the basic contact imaging mode, a sharp tip mounted on a flexible cantilever is scanned by a piezoelectric ceramic over the sample surface (Fig. 7.1). The deflection of the cantilever is continuously monitored through an optical lever (laser and photodiode system) that records vertical and lateral deflection. In the constant force mode, a feedback loop acts on the piezoelectric ceramic to maintain a constant cantilever height, and thus the applied force on the sample is kept constant. In tapping mode, the cantilever is oscillated near its resonant frequency while scanning over the surface. In this mode, the contact between the tip and the sample is defined as a decrease in the resonance amplitude, and the feedback loop adjusts the cantilever height in order to keep the amplitude constant. These two basic imaging modes result in topographic images of the samples recorded line by line while raster scanning the sample with the tip. The AFM, however, is much more than a simple imaging tool as it is also able to detect forces as small as a few piconewtons. In the force spectroscopy (FS) mode (Fig. 7.1), the tip is no longer scanned over the surface, but rather approaches and retracts from the surface. FS results in force *versus* distance curves that can be analyzed in terms of contact point, nanomechanical properties, and adhesion forces. Moreover, it is also possible to record force curves according to a predefined matrix, resulting in height maps, nanomechanical maps, or adhesion maps acquired point-by-point with a force curve at each point. The latter mode is named force volume and has now evolved toward modes that record force–distance curves at very high speeds (Dufrêne et al. 2013; Chopinet et al. 2013).

In this chapter, we describe how AFM can refine our understanding of the yeast cell wall. Firstly, we discuss immobilization methods for yeasts so that they can withstand the tip lateral forces induced during scanning. Immobilization methods must not modify the yeast cell wall. This is an essential problem that needs to be solved properly before any AFM experiments can be pursued. Next, we focus on imaging data to demonstrate interest in live-cell AFM. In particular, we show that AFM is capable of recording cell growth at the single-cell level, as it can work in liquid (culture broth) and at a controlled temperature (microorganism's growth temperature). Then, we examine the measurement of nanomechanical properties of the yeast cell wall and attempt to make a link between the biochemical composition of the cell wall and its nanomechanical properties. We also examine how drugs modify the nanomechanical properties of the yeast cell wall. Finally, we look at single-molecule and single-cell experiments. The prerequisite for such experiments is functionalization of the AFM tip with a molecule of interest or with a cell. We describe these functionalization methods followed by some interesting results of single-molecule or single-cell FS experiments.

7.2 Immobilization of Yeast Cells

AFM is a scanning probe technique and therefore the tip when scanning induces lateral forces on the sample. If the sample (in our case yeast cells) is not properly immobilized, no images and no force curves can be recorded. The first attempt to achieve a firm immobilization of microorganisms, in general, was to fix the cells by air drying (Canetta et al. 2006) or by chemical fixation (Louise Meyer et al. 2010). However, these methods surely induce cell wall modifications. Other strategies were developed to overcome this difficulty; cells were immobilized in gelatin (Gad and Ikai 1995) or trapped into the pores of polycarbonate filters (Touhami et al. 2003a). These techniques have been widely used over the recent years (Francius et al. 2008; Alsteens et al. 2008; Dague et al. 2008b; Gilbert et al. 2007); however, it can lead to tip pollution in the case of gelatin trapping, or it can submit cells to mechanical forces in the case of cells trapped in pores. Also, both techniques are time-consuming because cells are spread all over the sample and can be quite difficult to find. To circumvent these problems, recent developments were dedicated to the fabrication of polydimethylsiloxane (PDMS) stamps structured at the microscale with wells of different sizes adapted to yeast cell diameter (Dague et al. 2011; Formosa et al. 2015a). Then, the cells are assembled into the microwells using the convective/capillary deposition technique, as shown in Fig. 7.2a. This results in arrays of living cells, immobilized

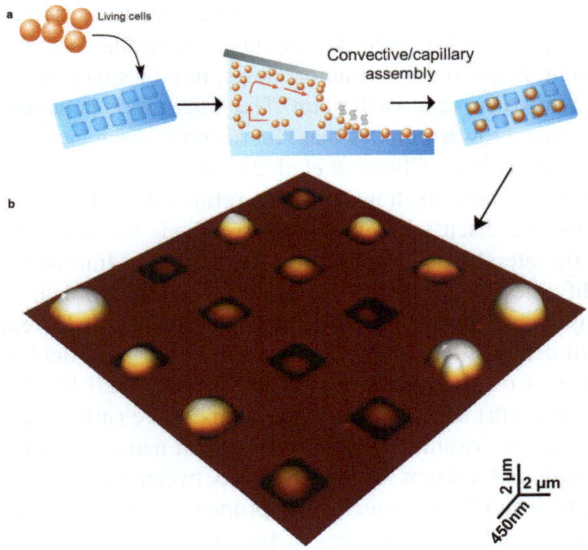

Fig. 7.2 Assembling yeast cells in a polydimethylsiloxane (PDMS) stamp using convective capillary deposition. **a** Principle of convective capillary deposition used to organize the yeast cells in the holes of a PDMS stamp. A meniscus is created between a glass slide and the microstructured stamp. Evaporation at the meniscus creates a convective flux in the liquid that concentrates the cells in the meniscus. The slide is then pulled over the surface. When encountering a hole, the meniscus is caught and a single cell is immobilized in the hole. **b** Atomic force microscopy (AFM) 3D height image of an array of 16 cells organized in a PDMS stamp

in a defined place. Figure 7.2b shows an arrangement of 16 cells organized in 16 holes. This innovative immobilization method makes it possible to analyze many more cells than in the past, which will result in an increase in the statistical meaning of AFM results.

7.3 Imaging Living Yeast Cells by Atomic Force Microscopy

Having overcome the immobilization issue, live, unmodified cells can be imaged at high resolution using AFM. Imaging living microorganisms at high resolution is clearly a challenge. Optical microscopy has been limited for a long time by diffraction (Chaps. 1–4), and its variation now allows sub-diffraction imaging (Jensen and Crossman 2014), while electron microscopy (Chaps. 6, 8) usually requires working in a vacuum and thus is not capable of imaging live cells. In this context, AFM provides the unique opportunity to image yeasts, neither with photons nor electrons, but with a sharp tip in growth media with temperature control and at high resolution (Dupres et al. 2009). The first images of living *S. cerevisiae* cells, immobilized in

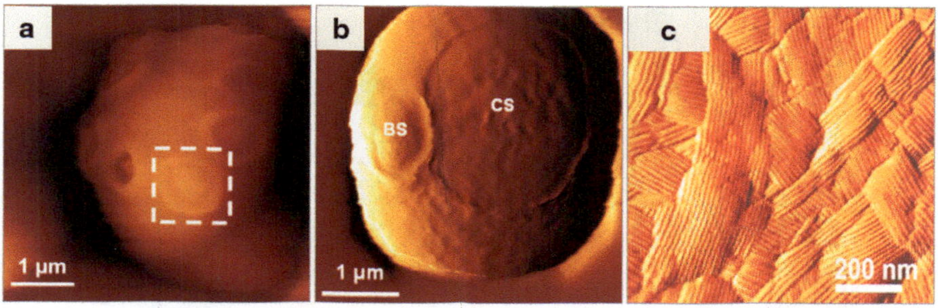

Fig. 7.3 Atomic force microscopy (AFM) high-resolution imaging of *Saccharomyces cerevisiae* with a ring made of eight bud scars *(BSs)* surrounding the cell. The *dotted line square* is centered on a *BS* (**a**), the circular structure *(CS)* induced by heat shock on *S. cerevisiae*, next to a *BS* (**b**), and the rodlet layer made of hydrophobins of *Aspergillus fumigatus* (**c**)

gelatin, showed bud and birth scars (Kasas and Ikai 1995). Figure 7.3a presents an example of a mother cell of *S. cerevisiae* with multiple bud scars (BSs), one of which is surrounded by a dashed square (Pillet et al. 2014b). Such features at the surface of yeast cells have been described in multiple studies (Touhami et al. 2003a; Dague et al. 2010; Alsteens et al. 2008; Adya et al. 2006) by different authors. In Fig. 7.3b, a BS can be seen next to a different and larger structure, named by the authors of this study a circular structure (CS, Pillet et al. 2014b), that appears on the *S. cerevisiae* surface after an hour of heat shock at 42 °C. This circular ring reaches 3 µm in diameter and is initiated on a BS. The authors showed that the formation of this structure required a functional budding process, as no CSs were observed on yeasts in which genes involved in the budding process were deleted.

At the surface of fungal spores, such as spores of *Aspergillus fumigatus,* proteins named hydrophobins are auto-organized and confer hydrophobicity to the conidia (Fig. 7.3c). These proteins create a rodlet layer (Beever and Dempsey 1978; Aimanianda et al. 2009) that can be imaged by AFM at high resolution (Dufrêne et al. 1999; Dague et al. 2008a; Zykwinska et al. 2014; Ma et al. 2005). Hydrophobins are amyloid proteins that are assembled into fibrils spaced from each other with 10 nm. This rodlet layer is also described on spores of bacteria such as *Bacillus atropheus* (Plomp et al. 2007a) or *Clostridium novyi* (Plomp et al. 2007b). The disruption of this layer during spore germination has been followed in real time (Dague et al. 2008c) which demonstrated that AFM was able to image dynamic processes such as germination on living cells. After 1 h of germination, the rodlet layer is slightly disrupted. After 2 h, the spore surface is heterogeneous; some regions still have an altered rodlet layer, whereas other regions present an amorphous surface. Finally, after 3 h, the whole surface of the spore is amorphous. Unfortunately, it was impossible to image the emission of a germ tube, probably because the immobilization of the conidia in the pore of a polycarbonate filter made that impossible. However, measures directly performed on small areas of hyphae were possible. The Dahms' lab demonstrates, using this method, that galactofuranose is a key constituent for the organization of the *Aspergillus nidulans* hyphal wall (Paul et al. 2011).

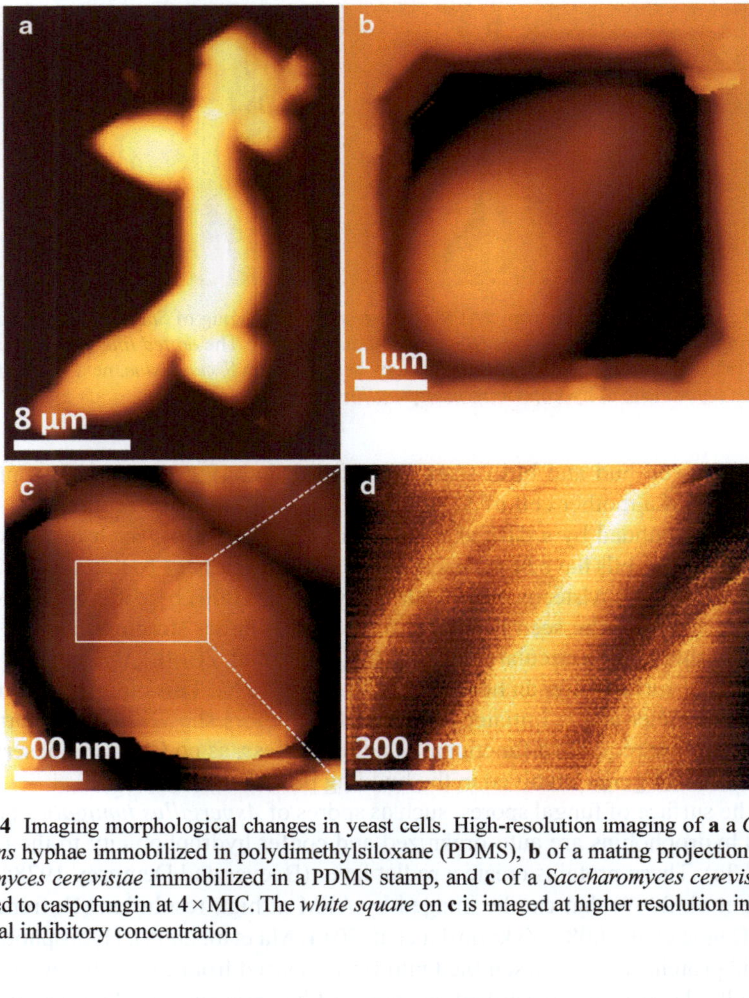

Fig. 7.4 Imaging morphological changes in yeast cells. High-resolution imaging of **a** a *Candida albicans* hyphae immobilized in polydimethylsiloxane (PDMS), **b** of a mating projection of *Saccharomyces cerevisiae* immobilized in a PDMS stamp, and **c** of a *Saccharomyces cerevisiae* cell exposed to caspofungin at 4 × MIC. The *white square* on **c** is imaged at higher resolution in **d**. *MIC* minimal inhibitory concentration

AFM also allowed the direct imaging of *C. albicans* hyphal growth, as shown in Fig. 7.4a (Formosa et al., personal communication), which follows from studies in which both fixed and live/growing hyphae were imaged by AFM (Ma et al. 2005, 2006). In Fig. 7.4a, the ramifications of the fungal hyphae are clearly observed. Hyphal forms of *C. albicans* are being used by the fungus to invade host tissues; the possibility to study them by AFM is therefore relevant for clinical microbiologists. It opens the doors to fundamental questions on the formation of these hyphae; for example, mutated strains for surface proteins could present modified growth processes, which could be monitored by AFM. Also, their interactions with host cells can be probed using AFM; for example, a recent study by El-Kirat-Chatel et al. focused on the interaction of *C. albicans* hyphae with macrophages and showed the ability of the fungus to grow hyphae directly into the macrophagic cells that resist them (El-Kirat-Chatel and Dufrêne 2012). Another phenomenon that can be imaged, thanks this time to the immobilization method in PDMS stamps, is the emission of mating projection by *S. cerevisiae* exposed to the α-factor (Fig. 7.4b). The α-factor, a yeast sexual hormone, triggers the formation of characteristic mating projections, also named "shmoos," by haploid yeast strains of a mating type (Merlini et al. 2013). This is an important process that facilitates contact between two partner cells so that they can fuse to form a diploid zygote. The study of yeast mating has many implications (e.g., in understanding analogous fundamental biological processes in higher eukaryotes), and its study by AFM could give new insights into the mechanisms underlying this process. For example, different genes could be implicated in this process using mutated strains in the presence of α-factor. However, many other types of experiments, including experiments with functionalized AFM tips recognizing specific proteins at the tip of shmoos or on the cell surface, can also be envisioned, knowing the great possibilities of such a technology.

Finally, AFM also allows to image changes in surface morphology induced by drugs such as antifungals. Figure 7.4c, d shows an *S. cerevisiae* cell treated with a high dose (4 × minimal inhibitory concentration) of caspofungin, an antifungal drug used for fungal infections (Denning 2003; Formosa et al. 2013). Upon treatment, the cell is no longer spherical but is elongated, resembling *Schizosaccharomyces pombe* cells. The antifungal-treated cell also presents a surprising feature on its surface (Fig. 7.3d), with rings up to 15 nm high, indicating altered cell division likely associated with impairment of cytokinesis.

Taken together, these examples show the importance of AFM, which offers the possibility to work on live yeast cells in liquid conditions. Indeed, AFM can be used to image yeast cell morphologies at the nanoscale, such as hyphae or shmoos, the ultrastructures naturally present at their surface (bud and birth scars), or those induced by stress (e.g., a protease; Ahimou et al. 2003) or the circular structure induced by thermal stress and rings induced by an antifungal treatment.

7.4 Probing the Nanomechanical Properties of Yeast Cells

To probe the nanomechanical properties of living yeasts, such as elasticity, spring constant, or turgor pressure, AFM is used in the FS mode. The nanomechanical property most often used to describe the cell wall of yeasts is elasticity; therefore, we focus on this parameter in this chapter. The elasticity can be deduced from the approach force–distance curves obtained in the FS mode. Force–distance curves can be converted into indentation curves, which are then fitted through the Hertz model, to extract the Young modulus value, meaning the elasticity, expressed in pascals (Hertz 1881). The elasticity of the cell wall reflects its composition but also its molecular organization. In 2003, Touhami and coworkers (Touhami et al. 2003b) mapped the nanoscale elasticity of *S. cerevisiae*. Specifically, they compared the nanomechanical properties of a BS with the rest of the cell wall. They mapped a higher resistance on the BS than on the rest of the cell and correlated this result with the increased amount of chitin in the BS. A few years later, this result was reproduced on *S. cerevisiae* but not for another yeast species, *Saccharomyces carlbergensis* (Alsteens et al. 2008). This result demonstrated that the correlation between nanomechanical data and the cell wall composition is not always straightforward.

Indeed, different components of the yeast cell wall are interconnected to form macromolecular complexes (Orlean 2012; Free 2013), which can be modified upon stress or if genes involved in cell wall synthesis are missing. A recent study showing the differences that can take place in the nanomechanical properties of yeast cells focused on yeast mutants of *S. cerevisiae* defective in cell wall architecture (Dague et al. 2010). In this work, the authors showed that native wild-type cells of *S. cerevisiae* had a global cell wall elasticity of 1.6 MPa, whereas its isogenic mutants defective in enzymes involved in cell wall cross-linking and assembly (*gas1, chr1chr2* mutants), or with a reduction of their chitin content (*chs3* mutant), had a reduced Young's modulus compared to wild-type cells. However, mutants with reduced β-glucan (*fks1*), mannan *(mnn9)* content, or defective in the regulation of the cell wall biosynthesis *(knr4)* presented higher Young's modulus values compared to wild-type cells. Thus, these results show that the nanomechanical properties of yeast cells are dependent not only on the composition of the cell wall but also on the intrinsic molecular organization of the cell wall, as shown by Paul et al. (2011).

The relationship between elasticity of the yeast cell wall and its composition/molecular organization has also been observed in a different context—thermal stress (Pillet et al. 2014b). In this study, the authors showed that thermal stress induced an increase in the chitin content of the cell wall, which was accompanied by an increase in the Young's modulus values. When yeast cells are submitted to a parietal stress, one of the first defense mechanisms is the overproduction of chitin (Ram et al. 1998). Chitin, being a rigid polymer, increases the elasticity of the cell wall when overproduced. Finally, previous studies on the effects of caspofungin on the yeast cell wall of *S. cerevisiae* and *C. albicans* also showed modification of the viscoelastic properties of cells upon treatment with this antifungal (Formosa et al. 2013; El-Kirat-Chatel et al. 2013). Figure 7.5 presents nanoindentation measurements performed on cells of *C. albicans* in native conditions, or treated by two different doses of caspofungin (0.5 and 4 × MIC). Figure 7.5a, b, c depicts elasticity maps of the whole cells immobilized in PDMS stamps; Fig. 7.5d, e, f shows elasticity maps recorded on small areas of 1 μm^2 on top of the corresponding cells; and, finally, Fig. 7.5g, h, i shows the distributions of the Young's moduli obtained for each pixel on the local elasticity maps for which the darker red the pixel, the higher the Young's modulus. We can see clearly from this figure that treatment with caspofungin results in an increase in the Young's moduli. Remarkably, the higher the caspofungin dose, the higher the Young's moduli, which was correlated with higher chitin content in the cell wall of caspofungin-treated yeasts. This work shows the correlation between elasticity of the yeast cell wall and its composition.

These results show how AFM can be used as a force machine to probe the cell wall of yeast cells in their native state and when submitted to genetic stress (mutants) or external stresses (heat shock and caspofungin treatment). Altogether, these nanomechanical data give new insights into the yeast cell wall organization and remodeling in response to different types of stresses.

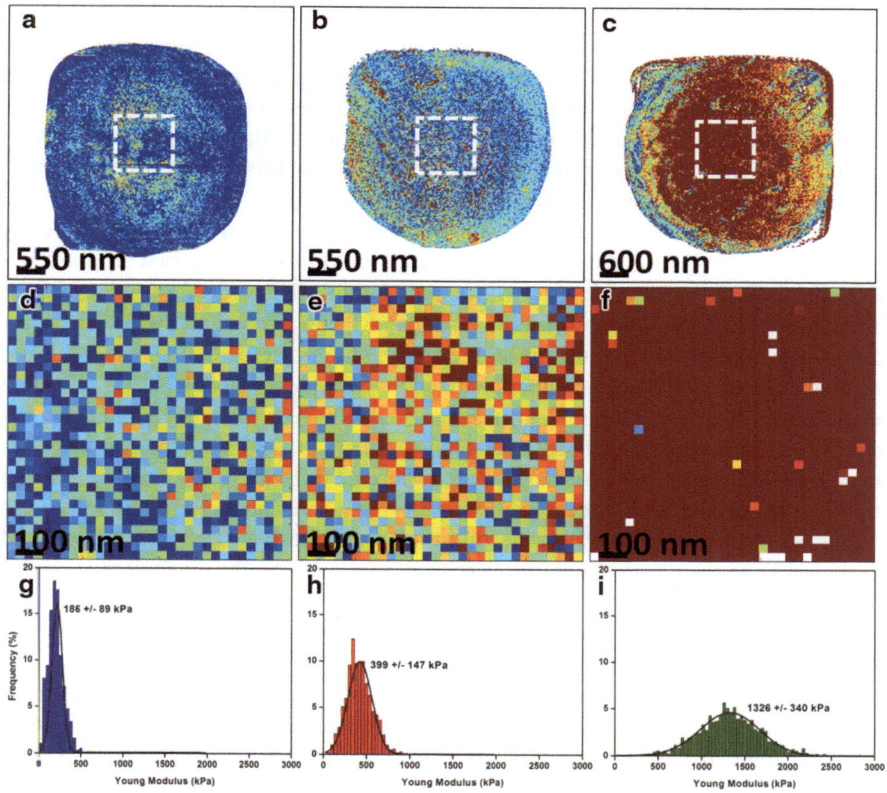

Fig. 7.5 Mapping of *C. albicans* cell surface elasticity. **a–c** Elasticity maps (z range = 0.5 MPa) of a native cell (**a**), of a cell treated with caspofungin at 0.5 × MIC (0.047 µg/ml) (**b**), and of a cell treated with caspofungin at 4 × MIC (0.376 µg/ml) (**c**). **d**, **e**, and **f** Local elasticity maps (z range = 0.5 MPa) recorded on a 1-µm area *(white dashed squares)* on the top of the cells in panels **a–c**, respectively. **g**, **h**, and **i** Distributions of Young's moduli ($n = 1024$) corresponding to the local elasticity maps in panels **d–f**, respectively. (Reprinted with permission from Formosa et al. 2013)

7.5 Single-Cell and Molecule Force Spectroscopy

Another property that can be probed by AFM is adhesion. Indeed, specific molecular interactions are the basis of various biochemical and biological processes. In order to gain significant data on these interactions, AFM tips can be functionalized with molecules that will interact specifically with target molecules at the surface of the cells. These experiments, performed with functionalized AFM tips, are called single-molecule force spectroscopy (SMFS) experiments. Of the various strategies to functionalize AFM tips with biomolecules, some consist of nonspecific adsorption of proteins (e.g., bovine serum albumin; BSA) to the silicon nitride surface of AFM tips (Florin et al. 1994), or the chemical fixation of biomolecules by sulfur–gold bonds to gold-coated AFM tips. This last strategy has been successfully used to functionalize AFM tips with methyl groups, CH_3, to probe the hydrophobic characteristics of the rodlet layer of *A. fumigatus* (Alsteens et al. 2007; Dague et al. 2007). In this study, the authors showed that hydrophobic tips enable quantification of surface hydrophobicity on live cell surfaces, and how this hydrophobicity relates to a function such as surface adhesion or drug interaction.

It is also possible to covalently link a molecule containing amino groups directly to the silicon nitride AFM tip. Toward this end, AFM tips must be first amino-functionalized either by esterification with ethanolamine (Hinterdorfer et al. 1996) or silanization with aminopropyl-triethoxysilane (APTES; Ros et al. 1998). Then, the amino-functionalized tip must be bridged to the biomolecule of interest, achieved through the use of hetero-bifunctionalized polyethylene glycol (PEG; Wildling et al. 2011; Ebner et al. 2008; Kamruzzahan et al. 2006) or an aldehyde–phosphorus dendrimer, as we previously described (Jauvert et al. 2012). The second strategy has been used to map the polysaccharides at the surface of living yeast cells, with AFM tips functionalized with concanavalin A, a protein that interacts specifically with carbohydrates (Gad et al. 1997). In this study, the adhesive forces were calculated from the AFM retract portion of force curves by measuring the piezoelectric retraction force required to break the interaction between the lectin and the recognized carbohydrate. Such measurements allowed the authors to conclude that mannans were uniformly distributed on the cell wall surface.

This functionalization strategy has also been used to map the surface properties of the pathogenic yeast *C. albicans*. This pathogenic yeast species has emerged as a major public health problem in the past two decades. This opportunistic pathogen causes a wide range of infections from surface to mucosal and bloodstream infections (Gow and Hube 2012). In order to colonize and subsequently disseminate in the bloodstream, *C. albicans* first needs to adhere to different biotic substrates. This first stage of infection (Naglik et al. 2011) is mediated by adhesins that are found on the surface of the yeast cell wall. Many of these adhesins are mannoproteins; among them, one identified as having a major role in host cell attachment is the agglutinin-like sequences (Als) family (Hoyer 2001). The Als were initially reported as having homology to the proteins responsible for autoagglutination in the baker's yeast *S. cerevisiae*. Eight *ALS* genes have been identified, and all are primarily involved in host–pathogen interactions (Hoyer et al. 2008). In a recent study by Beaussart et al., the authors used SMFS experiments to map the localization of Als3 on the surface of *C. albicans* hyphae (Beaussart et al. 2012) using an anti-Als3 antibody-functionalized tip to probe different parts of the germinating tube, that is, on the germinating yeast (Fig. 7.6a, b, c) and on the germ tube (Fig. 7.6d, e, f). The Als3 protein distribution was very different with the number and length of unfolding events higher on the germ tube, accompanied by longer unfoldings, indicating that the adhesin Als3 is much more exposed on the germ tube than on the germinating yeast. These results are consistent with the specific expression of the *ALS3* gene during the yeast–hyphae transition (Liu and Filler 2011).

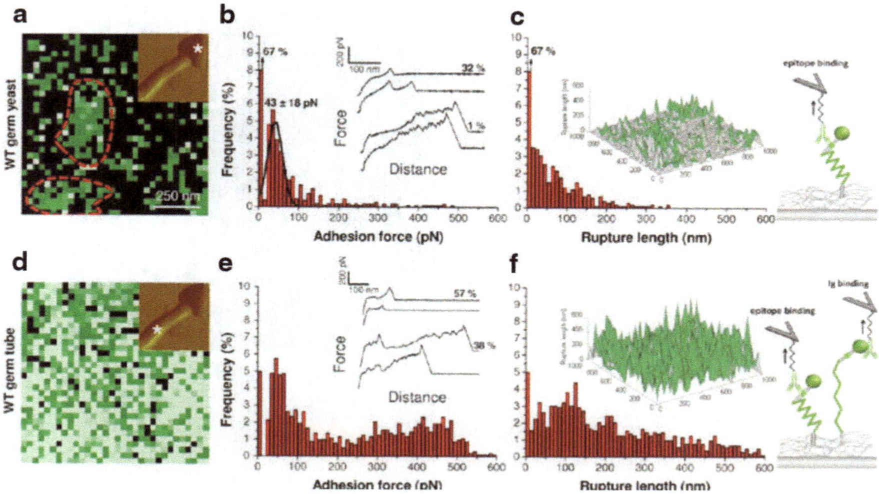

Fig. 7.6 Cellular morphogenesis leads to a major increase in the distribution and extension of Als3 proteins. **a**, **d** Adhesion force maps (1 μm^2, z range = 300 pN) recorded in buffer on the yeast (**a**) and germ tube (**d**) of a germinating cell using an anti-Als3 tip. *Insets:* deflection images in which the *asterisk* indicates where the force maps were recorded. The *dashed lines* in **a** emphasize Als3 clusters. **b**, **e** Corresponding adhesion force histograms (n=1024) together with representative force curves. **c**, **f** Histograms of rupture distances (n=1024) and 3D-reconstructed polymer maps (false colors, adhesion forces in *green*). The data are representative of several independent experiments using different tip preparations and cell cultures. (Reprinted with permission from Beaussart et al. 2012)

Another characteristic of the Als proteins is the presence of an amyloid-forming sequence, which enables their aggregation, under certain conditions, into amyloids at the surface of *C. albicans* cells. A recent study by Alsteens et al. showed that the formation and propagation of Als5 protein nanodomains of the Als5 protein at the surface of *S. cerevisiae* yeast cells overexpressing Als5 was force induced (Alsteens et al. 2010). AFM tips functionalized with an anti-Als5 antibody showed that a localized delivery of piconewton forces by the AFM tip could initiate the formation and propagation of Als5 nanodomains over the cell (Fig. 7.7a, b). Thus, this process may be involved in cellular adhesion, in response to mechanical stimuli—important processes for the early stages of *C. albicans* infection.

The clustering of adhesins at the yeast surface has also been described in an industrial nonpathogenic strain of *S. cerevisiae* (Schiavone et al. 2015) (Fig. 7.7c, d). A recent study by Schiavone et al. used a transcriptomic analysis to identify the protein responsible for the adhesive nanodomains. Their results showed that the strain producing adhesive nanodomains also overexpressed the Flo11 protein, an adhesin similar to the Als protein family that also contains an amyloid sequence. Finally, the use of Quantitative Imaging™ (an advanced force-volume-based mode) made it possible to map, with unprecedented resolution, adhesive nanodomains at the *C. albicans* cell surface (Fig. 7.7e, f; Formosa et al. 2015b). In this work by Formosa

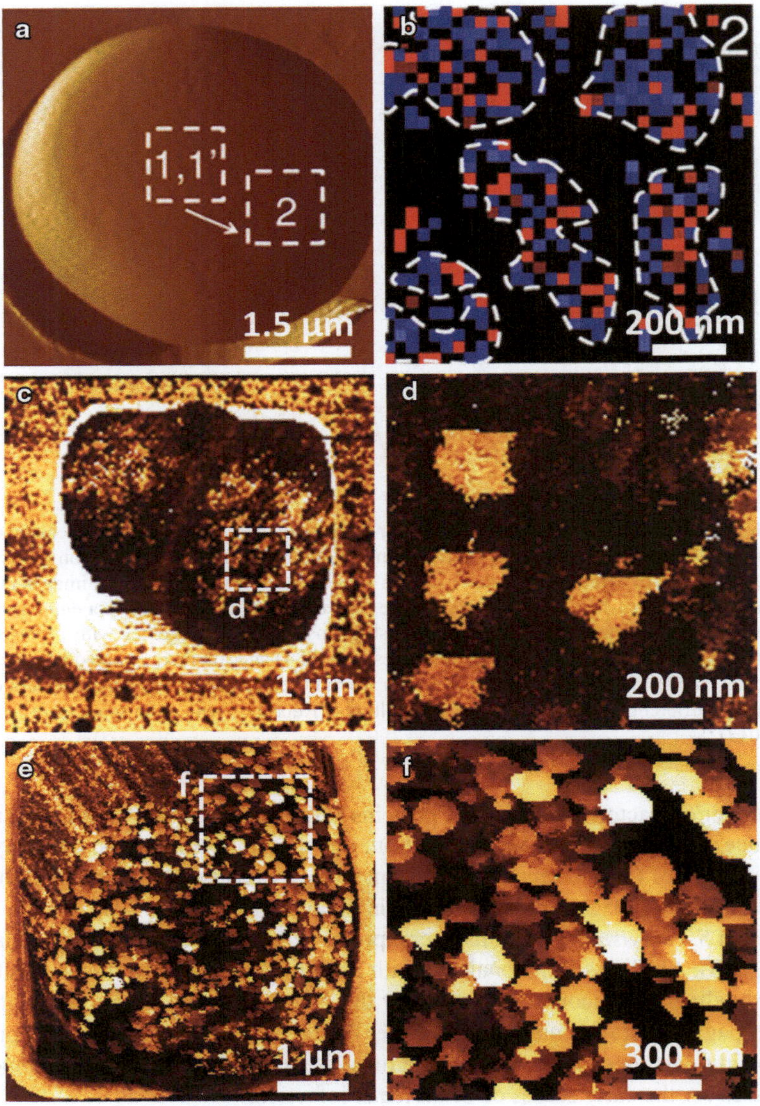

Fig. 7.7 Adhesive nanodomains at the yeast surface. **a** Atomic force microscopy (AFM) height image of *S. cerevisiae* trapped in a pore of a polycarbonate membrane. **b** Adhesion map, recorded in the *dotted square 2,* after force inducing the clustering in area *1,1'*, showing the aggregation in nanodomains of the Als5 protein. *Blue* and *red pixels* correspond to forces smaller and larger than 150 pN, respectively. **c** AFM adhesion map of an *S. cerevisiae* cell immobilized in a microstructured polydimethylsiloxane (PDMS) stamp. **d** Adhesion map corresponding to the *dotted square d.* The adhesion forces range from 0 to 2 nN. **e** Adhesion map recorded on a *C. albicans* cell immobilized in a microstructured PDMS stamp. **f** Adhesion map corresponding to the *dotted square f.* Adhesion forces range from 0 to 2 nN

et al., nanodomain size, stiffness, and adhesive properties were characterized, with nanodomains divided into two classes. Some are stiff and behave like hydrophobic structures (high adhesion force), whereas others are softer and unfold in a manner similar to proteins. The first class of nanodomains was interpreted as being amyloid plaques and the second class as protein aggregates. Altogether, the formation of adhesive nanodomains at the yeast cell surface seems to be a general phenomenon; further work is required to understand what triggers their formation, if and how they are used by cells to adhere to or infect hosts.

With these last examples, we show that FS in conjunction with functionalized AFM tips is a powerful tool to better understand the molecular adhesive properties of living pathogenic yeast cells. These data will eventually contribute to preventing *C. albicans* infections, as well as in the identification of new potential targets for antifungal drugs.

Since it is possible to functionalize an AFM tip with a molecule, it is also possible to functionalize an AFM tip with a living cell. This type of experiment is called single-cell force spectroscopy (SCFS) experiment. The principle is the same as for SMFS, but this time a living cell is immobilized onto the cantilever becoming the AFM tip, which is engaged to interact with the surface or with another cell (Bowen et al. 2000, 2001). This strategy is reemerging, and there are only few papers in which it is used for the study of living yeast cells (Bowen et al. 2000, 2001; Stewart et al. 2013). However, Dufrêne's group has recently published a study that aims to measure the adhesive forces between a yeast cell immobilized on a cantilever and a hyphae of the same yeast species *C. albicans* (Alsteens et al. 2013). They show that the *C. albicans* adhesin, Als3, on the hypal surface was responsible for their adhesion with yeast cells. In the context of biofilm formation, these new data offer the first new insight into the interactions between two morphotypes of the same yeast species. The same team also used SCFS to quantify adhesion between cells of *C. albicans* and another type of pathogenic microorganism, *Staphylococcus epidermidis* (Beaussart et al. 2013). Indeed, these two types of microorganisms are often colocalized in human infections (Peleg et al. 2010). The results of this study showed that the fungal molecules involved in the interactions with *S. epidermidis* were adhesins, once again, as well as 0-mannosylations. It is interesting to note that, using the same cell probe strategy, the interactions between vaginal staphylococci and lactococci were probed (Younes et al. 2012). This work, pioneering in the field of microbiology, opens doors to new questions about fungal adhesion, an important process involved in the first stages of human yeast infections. Such studies could lead, for example, to the identification of anti-adhesion drugs that would prevent coinfections by *C. albicans* and *S. epidermidis*. The lack of statistical data has been recently addressed, and promising perspectives are evolving from the work of Potthoff (Potthoff et al. 2012), which aims to serially quantify adhesion forces using yeast probes.

7.6 Conclusions

AFM has emerged as a significant technology in the life sciences over the past 20 years. The number of publications using AFM is exponentially growing (Pillet et al. 2014a), even if the number of studies dedicated to living cells only slowly increases. The latter issue may relate to the difficulty of immobilizing live cells and avoiding damage to the cell membrane during immobilization. For yeast cells, physical trapping in pores of polycarbonate membranes or in holes of microstructured PDMS stamps is appropriate. As an imaging technology, AFM helps refine our understanding of structures exposed on the fungal cell wall, such as BSs and rodlet layers. Interestingly, AFM is also able to track morphological modifications during cell growth. More than an imaging technology, AFM is also a force machine able to probe the cell nanomechanical properties or adhesion. The data collected from FS experiments have broadened our perception of the fungal cell wall. The well-known biochemical composition of the cell wall covers up the extremely complex and dynamic interplay of its organization and architecture. Therefore, much more work is required for a fully comprehensive description of fungal cell walls and the application of that knowledge to relevant issues in the life sciences.

References

Adya AK, Canetta E, Walker GM (2006) Atomic force microscopic study of the influence of physical stresses on *Saccharomyces cerevisiae* and *Schizosaccharomyces pombe*. FEMS Yeast Res 6:120–128

Ahimou F, Touhami A, Dufrêne YF (2003) Real-time imaging of the surface topography of living yeast cells by atomic force microscopy. Yeast Chichester Engl 20:25–30

Aimanianda V et al (2009) Surface hydrophobin prevents immune recognition of airborne fungal spores. Nature 460:1117–1121

Alsteens D, Dague E, Rouxhet PG, Baulard AR, Dufrêne YF (2007) Direct Measurement of hydrophobic forces on cell surfaces using AFM. Langmuir 23:11977–11979

Alsteens D et al (2008) Structure, cell wall elasticity and polysaccharide properties of living yeast cells, as probed by AFM. Nanotechnology 19:384005

Alsteens D, Garcia MC, Lipke PN, Dufrêne YF (2010) Force-induced formation and propagation of adhesion nanodomains in living fungal cells. Proc Natl Acad Sci U S A 107:20744–20749

Alsteens D, Van Dijck P, Lipke PN, Dufrêne YF (2013) Quantifying the forces driving cell–cell adhesion in a fungal pathogen. Langmuir 29:13473–13480

Beaussart A et al (2012) Single-molecule imaging and functional analysis of als adhesins and mannans during *Candida albicans* morphogenesis. ACS Nano 6:10950–10964

Beaussart A et al (2013) Single-cell force spectroscopy of the medically important *Staphylococcus epidermidis-Candida albicans* interaction. Nanoscale 5:10894–10900

Beever RE, Dempsey GP (1978) Function of rodlets on the surface of fungal spores. Nature 272:608–610

Binnig G, Quate CF (1986) Atomic force microscope. Phys Rev Lett 56:930–933

Bowen WR, Lovitt RW, Wright CJ (2000) Direct quantification of *Aspergillus niger* spore adhesion to mica in air using an atomic force microscope. Colloid Surf Physicochem Eng Asp 173:205–210

Bowen WR, Lovitt RW, Wright CJ (2001) Atomic force microscopy study of the adhesion of *Saccharomyces cerevisiae*. J Colloid Interface Sci 237:54–61

Canetta E, Adya AK, Walker GM (2006) Atomic force microscopic study of the effects of ethanol on yeast cell surface morphology. FEMS Microbiol Lett 255:308–315

Chaffin WL (2008) *Candida albicans* cell wall proteins. Microbiol Mol Biol Rev 72:495–544

Chopinet L, Formosa C, Rols MP, Duval RE, Dague E (2013) Imaging living cells surface and quantifying its properties at high resolution using AFM in QI™ mode. Micron 48:26–33

Dague E et al (2007) Chemical force microscopy of single live cells. Nano Lett 7:3026–3030

Dague E, Delcorte A, Latge JP, Dufrene YF (2008a) Combined use of atomic force microscopy, X-ray photoelectron spectroscopy, and secondary ion mass spectrometry for cell surface analysis. Langmuir 24:2955–2959

Dague E, Alsteens D, Latgé J-P, Dufrêne YF (2008b) High-resolution cell surface dynamics of germinating *Aspergillus fumigatus* conidia. Biophys J 94:656–660

Dague E, Alsteens D, Latge J-P, Dufrene Y (2008c) High-resolution cell surface dynamics of germinating *Aspergillus fumigatus* conidia. Biophys J 94:1–5

Dague E, Bittar R, Durand F, Martin-Hyken H, François JM (2010) An atomic force microscopy analysis of yeast mutants defective in cell wall architecture. Yeast 27:673–784

Dague E et al (2011) Assembly of live micro-organisms on microstructured PDMS stamps by convective/capillary deposition for AFM bio-experiments. Nanotechnology 22:395102

Denning DW (2003) Echinocandin antifungal drugs. Lancet 362:1142–1151

Dufrêne YF, Boonaert CJP, Gerin PA, Asther M, Rouxhet PG (1999) Direct probing of the surface ultrastructure and molecular interactions of dormant and germinating spores of Phanerochaete chrysosporium. J Bacteriol 181:5350–5354

Dufrêne YF, Martínez-Martín D, Medalsy I, Alsteens D, Müller DJ (2013) Multiparametric imaging of biological systems by force-distance curve-based AFM. Nat Methods 10:847–854

Dupres V, Alsteens D, Pauwels K, Dufrêne YF (2009) In vivo imaging of S-layer nanoarrays on corynebacterium glutamicum. Langmuir 25:9653–9655

Ebner A et al (2008) Functionalization of probe tips and supports for single-molecule recognition force microscopy. Top Curr Chem 285:29–76

El-Kirat-Chatel S, Dufrêne YF (2012) Nanoscale imaging of the *Candida*–macrophage interaction using correlated fluorescence-atomic force microscopy. ACS Nano 6:10792–10799

El-Kirat-Chatel S et al (2013) Nanoscale analysis of caspofungin-induced cell surface remodelling in *Candida albicans*. Nanoscale 5:1105–1115

Florin EL, Moy VT, Gaub HE (1994) Adhesion forces between individual ligand-receptor pairs. Science 264:415–417

Formosa C et al (2013) Nanoscale effects of caspofungin against two yeast species, *Saccharomyces cerevisiae* and *Candida albicans*. Antimicrob Agents Chemother 57:3498–3506

Formosa C, Pillet F, Schiavone M, Duval RE, Ressier L, Dague E (2015a) Generating living cells arrays for Atomic Force Microscopy studies. Nat Protoc 10(1):199–204

Formosa C, Schiavone M, Boisrame A, Richard ML, Duval RE, Dague E (2015b) Multiparametric imaging of adhesive nanodomains at the surface of Candida albicans by Atomic Force Microscopy. Nanomedicine NBM 11:57–65

Francius G, Domenech O, Mingeot-Leclercq MP, Dufrêne YF (2008) Direct observation of *Staphylococcus aureus* cell wall digestion by lysostaphin. J Bacteriol 190:7904–7909

Free SJ (2013) Fungal Cell Wall Organization and Biosynthesis. In: Friedman T, Dunlap JC, Goodwin SF (eds) Advances in genetics, vol 81. Academic Press, pp 33–82

Gad M, Ikai A (1995) Method for immobilizing microbial cells on gel surface for dynamic AFM studies. Biophys J 69:2226–2233

Gad M, Itoh A, Ikai A (1997) Mapping cell wall polysaccharides of living microbial cells using atomic force microscopy. Cell Biol Int 21:697–706

Gilbert Y et al (2007) Single-molecule force spectroscopy and imaging of the vancomycin/d-Ala-d-Ala interaction. Nano Lett 7:796–801

Gow NA, Hube B (2012) Importance of the *Candida albicans* cell wall during commensalism and infection. Curr Opin Microbiol 15:406–412

Hertz H (1881) Ueber die Berührung fester elastischer Körper. J Reine Angew Math 92:156–171

Hinterdorfer P, Baumgartner W, Gruber HJ, Schilcher K, Schindler H (1996) Detection and localization of individual antibody-antigen recognition events by atomic force microscopy. Proc Natl Acad Sci U S A 93:3477–3481

Hoyer LL (2001) The ALS gene family of *Candida albicans*. Trends Microbiol 9:176–180
Hoyer LL, Green CB, Oh S-H, Zhao X (2008) Discovering the secrets of the *Candida albicans* agglutinin-like sequence (ALS) gene family—a sticky pursuit. Med Mycol 46:1–15
Jauvert E et al (2012) Probing single molecule interactions by AFM using bio-functionalized dendritips. Sens Actuators B Chem 168:436–441
Jensen E, Crossman DJ (2014) Technical review: types of imaging-direct STORM. Anat Rec Hoboken NJ. doi:10.1002/ar.22960
Kamruzzahan ASM et al (2006) Antibody linking to atomic force microscope tips via disulfide bond formation. Bioconjug Chem 17:1473–1481
Kasas S, Ikai A (1995) A method for anchoring round shaped cells for atomic force microscope imaging. Biophys J 68:1678–1680
Lipke PN, Ovalle R (1998) Cell wall architecture in yeast: new structure and new challenges. J Bacteriol 180:3735–3740
Liu Y, Filler SG (2011) *Candida albicans* Als3, a multifunctional adhesin and invasin. Eukaryot Cell 10:168–173
Louise Meyer R et al (2010) Immobilisation of living bacteria for AFM imaging under physiological conditions. Ultramicroscopy 110:1349–1357
Ma H, Snook LA, Kaminskyj SGW, Dahms TES (2005) Surface ultrastructure and elasticity in growing tips and mature regions of *Aspergillus* hyphae describe wall maturation. Microbiology 151:3679–3688
Ma H, Snook LA, Tian C, Kaminskyj SGW, Dahms TES (2006) Fungal surface remodelling visualized by atomic force microscopy. Mycol Res 110:879–886
Merlini L, Dudin O, Martin SG (2013) Mate and fuse: how yeast cells do it. Open Biol 3:130008
Naglik JR, Moyes DL, Wächtler B, Hube B (2011) *Candida albicans* interactions with epithelial cells and mucosal immunity. Microbes Infect 13:963–976
Orlean P (2012) Architecture and Biosynthesis of the *Saccharomyces cerevisiae* Cell Wall. Genetics 192:775–818
Paul BC, El-Ganiny AM, Abbas M, Kaminskyj SGW, Dahms TES (2011) Quantifying the importance of galactofuranose in *Aspergillus nidulans* hyphal wall surface organization by atomic force microscopy. Eukaryot Cell 10:646–653
Peleg AY, Hogan DA, Mylonakis E (2010) Medically important bacterial–fungal interactions. Nat Rev Micro 8:340–349
Pillet F, Chopinet L, Formosa C, Dague É (2014a) Atomic force microscopy and pharmacology: from microbiology to cancerology. Biochim Biophys Acta (BBA—Gen Subj) 1840:1028–1050
Pillet F et al (2014b) Uncovering by atomic force microscopy of an original circular structure at the yeast cell surface in response to heat shock. BMC Biol 12:6
Plomp M, Leighton TJ, Wheeler KE, Hill HD, Malkin AJ (2007a) In vitro high-resolution structural dynamics of single germinating bacterial spores. PNAS 104:9644–9649
Plomp M et al (2007b) Spore coat architecture of *Clostridium novyi* NT spores. J Bacteriol 189:6457–6468
Potthoff E et al (2012) Rapid and serial quantification of adhesion forces of yeast and mammalian cells. PLoS ONE 7:e52712
Ram AFJ et al (1998) Loss of the plasma membrane-bound protein Gas1p in *Saccharomyces cerevisiae* results in the release of b1,3-Glucan into the medium and induces a compensation mechanism to ensure cell wall integrity. J Bacteriol 180:1418–1424
Ros R et al (1998) Antigen binding forces of individually addressed single-chain Fv antibody molecules. Proc Natl Acad Sci U S A 95:7402–7405
Schiavone M, Sieczkowski N, Castex M, Dague E, Marie François J (2015) Effects of the strain background and autolysis process on the composition and biophysical properties of the cell wall from two different industrial yeasts. FEMS Yeast Res 15(2) pii:fou012. doi:10.1093/femsyr/fou012
Stewart MP et al (2013) Wedged AFM-cantilevers for parallel plate cell mechanics. Methods San Diego Calif 60:186–194
Touhami A, Nysten B, Dufrêne YF (2003a) Nanoscale mapping of the elasticity of microbial cells by atomic force microscopy. Langmuir 19:4539

Touhami A, Nysten B, Dufrêne YF (2003b) Nanoscale mapping of the elasticity of microbial cells by atomic force microscopy. Langmuir 19:4539–4543

Wildling L et al (2011) Linking of sensor molecules with amino groups to amino-functionalized AFM tips. Bioconjug Chem 22:1239–1248

Younes JA, van der Mei HC, van den Heuvel E, Busscher HJ, Reid G (2012) Adhesion forces and coaggregation between vaginal staphylococci and lactobacilli. PLoS ONE 7:e36917

Zykwinska A, Pihet M, Radji S, Bouchara J-P, Cuenot S (2014) Self-assembly of proteins into a three-dimensional multilayer system: investigation of the surface of the human fungal pathogen Aspergillus fumigatus. Biochim Biophys (Acta BBA—Proteins Proteomics). doi:10.1016/j.bbapap.2014.03.001

Chapter 8
Future Directions in Advanced Mycological Microscopy

Kirk J. Czymmek and Tanya E. S. Dahms

8.1 Introduction

Recently, a truly remarkable series of innovative technical advances in cell biology tools have opened the door to new discoveries and ultimately a much deeper understanding of many biological phenomena. Indeed there are so many innovations, with often subtle variations, that it can be a significant challenge to identify which tool/approach is the most appropriate to address a scientific question. While the pace of progress is quite extraordinary, we welcome these advances that are quickly overcoming technological barriers/limitations of even just a few years ago. While these developments are exciting, this virtual renaissance for cell biologists and microscopists requires us to keep an ever-vigilant eye for the quick adoption of new strategies as appropriate. Fungal cell biologists are well positioned to leverage these advances due to the inherent microscopic dimensions of many fungal structures, suitable model organisms and their fundamental importance in our day-to-day life in relation to food, medicine, and disease.

In this chapter, our focus on future directions turned to select imaging topics that we believe have great potential to transform how we understand mycological research, including fungal interactions with the environment and other organisms. We not only place special emphasis on correlative microscopy (light, electron, and scanning probe), but also include the interrelated topics of advanced three-

K. J. Czymmek (✉)
North American Applications and Labs, Carl Zeiss Microscopy, LLC, One Zeiss Drive, Thornwood, NY 10594, USA
e-mail: Kirk.Czymmek@zeiss.com

T. E. S. Dahms (✉)
Department of Chemistry and Biochemistry, University of Regina, Regina, Saskatchewan S4P 1N7, Canada
e-mail: tanya.dahms@uregina.ca

T. E. S. Dahms and K. J. Czymmek (eds.), *Advanced Microscopy in Mycology*, Fungal Biology, DOI 10.1007/978-3-319-22437-4_8

dimensional (3D) electron microscopy and fluorescent biosensors. We consider all of these approaches naturally intertwined as key imaging building blocks that, when used together, significantly amplify the benefits and power of the individual tools.

8.2 Advances in the Application of Biosensors

As demonstrated throughout this book, fungal cell biology has benefited from an array of contrasting and complementary strategies that allow visualization of gross fungal structures, individual cells, subcellular compartments, and macromolecules. A great emphasis has been placed on understanding the polarized hyphal tip, and rightly so as it is the dominant form of fungal growth. Indeed, various fluorescent probes including small molecules (Kankanala et al. 2007; Hickey et al. 2004) and genetically encoded protein fusions (Czymmek et al. 2002, 2004, 2005) have been a boon to our understanding of tip growth. However, in fungi, fully realizing the significant progress demonstrated in plant and mammalian cell systems has been challenging, especially with genetically encoded biosensors for imaging ions, molecular interactions, and cell signaling dynamics. The vast majority of biosensors exploit Förster resonance energy transfer (FRET)-based folding/unfolding of a target protein that is a chimera with a pair of spectrally overlapping fluorescent proteins (FPs) to reveal dynamic and measurable responses. However, intensity and spectral changes (Miesenböck et al. 1998) and single fluorophore complementation strategies (Akerboom et al. 2012) can also be detected with single FP sensors. While no fundamental barriers prevent the application of genetically encoded biosensors in fungi, with numerous examples of standard FP fusions in the fungal kingdom (Chaps. 1–3; Czymmek et al. 2004), there are surprisingly few published examples. As such, we suggest that there are significant opportunities for new discoveries using biosensors to probe the inner workings of fungal cells. One conspicuous example is the ubiquitous calcium ion which as a universal secondary messenger in biological systems has been one of the best studied ions. However, visualizing calcium ion dynamics using organic fluorescent dyes in fungi has met with limited success, driving the use of FP-based calcium ion sensors, already widely used in animal and plant biology (Newman et al. 2011; Okumoto et al. 2012). Toward this end, the bioluminescent aequorin gene from the jellyfish *Aequoria victoria* was successfully expressed in fungi to determine cytosolic calcium concentration ($[Ca^{2+}]_c$) (Nelson et al. 2004). The study required significant effort to optimize codon usage but was rewarded with successful time-based calcium signaling, monitoring in populations of cells for the first time. Subsequently, successful expression and imaging of Cameleon YC3.60 (Nagai et al. 2004), the calmodulin-based $[Ca^{2+}]_c$ FRET sensor, in three economically important plant pathogenic fungi, *Magnaporthe oryzae, Fusarium oxysporum,* and *Fusarium graminearum* (Kim et al. 2012), was demonstrated. In this work, age/development dependent and pulsatile subcellular calcium dynamics and transient tip high $[Ca^{2+}]_c$ gradients were observed in various stages of vegetative growth and disease.

Genetic encoding can help circumvent the recalcitrant properties of the fungal cell and its tendency to sequester or otherwise compartmentalize/degrade exogenously introduced fluorescent probes, rendering them ineffective (Hickey et al. 2004). For those genetically tractable species, codon optimization and construct modification will likely be required along with species-dependent strategies. The array of biosensors already available and used in mammalian and plant cells can serve as a starting point for real-time monitoring of cellular activity. Those monitoring pH (Miesenböck et al. 1998), promotor activity (Terskikh et al. 2000), protein turnover (Subach et al. 2009), translocation (Varnai and Balla 2007), protease (Xu et al. 1998; Luo et al. 2001) calcium ions (Miyawaki et al. 1997; Kim et al. 2012; Akerboom et al. 2012), adenosine triphosphate (ATP; Imamura et al. 2009), cyclic adenosine monophosphate (cAMP; Zaccolo et al. 2000), cyclic guanosine monophosphate (cGMP; Honda et al. 2001; Sato et al. 2000), G-protein (Pertz et al. 2006) and kinase activation (Schleifenbaum et al. 2004), reactive oxygen species (ROS; Belousov et al. 2006), and membrane potential (Tsutsui et al. 2008) will allow many critical fungal processes to be explored in real time. Of course, many of these sensors can be applied in concert with other FP fusions and vital dyes that target organelles, cytoskeletal, and other proteins. We hope our readers will see the merit in exploiting biosensors in their own research projects and we encourage you to explore the possibilities (Reviewed in Frommer et al. 2009; Newman et al. 2011; Okumoto et al. 2012).

8.3 Three-Dimensional Electron Microscopy

Fungal structures, such as organelles, cells, mycelial networks/tissues, and fungal host/surface interactions, are inherently 3D. A far greater understanding of the structure–function relationship in these cells and tissues is now achievable with high-resolution correlation of chemical markers and structural components in the third dimension. Transmission electron microscopy (TEM), a mainstay in high-resolution fungal cell biology, especially when combined with freeze-substitution, antibody/affinity probe gold localization, and serial section analysis (Howard and Aist 1979), has tremendously impacted our understanding of hyphal tip architecture and function. TEM tomography, which typically generates tilt series (i.e., +/−70°) and creates 3D data sets from back projections (Harris et al. 2005; Hohmann-Marriott et al. 2006), provides unprecedented quantitative and structural perspectives of hyphal tip organization. While TEM is the gold standard for high-resolution ultrastructure, field emission scanning electron microscopy (FESEM) can now accommodate resin embedded samples that have been physically sliced into sections on a standard ultramicrotome mounted serially on tape, glass slide, or slotted grid, each substrate having unique benefits. Modern FESEMs are highly versatile and can image stained ultrathin sections with backscattered electron (BSE), secondary electron, and scanning transmission electron detectors yielding a TEM-like image. Alternatively, 3D tomograms of sliced volumes, much like TEM, can be automated

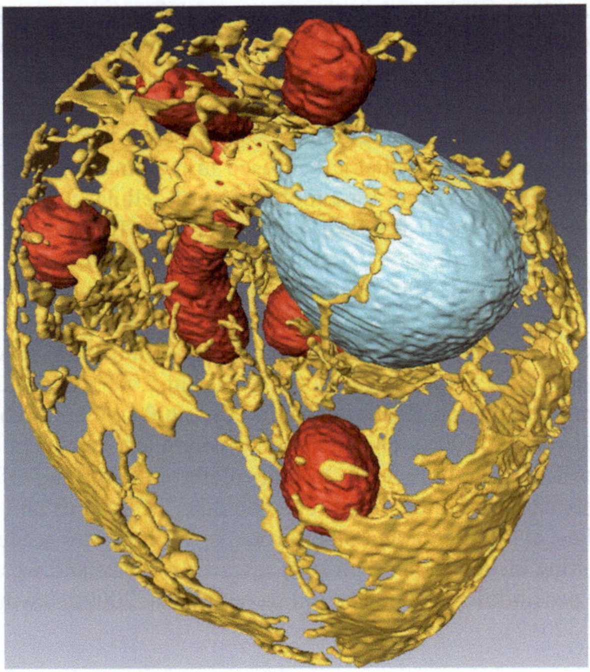

Fig. 8.1 High-resolution 3D reconstruction of a single cell of *Saccharomyces cerevisiae* acquired using focused ion beam scanning electron microscopy (*FIBSEM*; Chap. 6). The high-pressure frozen freeze-substituted, epoxy embedded yeast samples were acquired at 5 nm isotropic voxels. The *FIBSEM* approach enables extended 3D volumes of fungal structures, subcellular organelles and molecules at resolutions associated with electron microscopy. The nucleus is shown in *light blue,* mitochondria in *red,* and endoplasmic reticulum in *yellow*. Details of the protocols and findings can be found elsewhere (Wei et al. 2012). (Image provided courtesy of Jeff Caplan and Kirk Czymmek)

with serial block-face imaging (Helmstaedter et al. 2008) in which an ultramicrotome is either integrated directly into the electron microscope chamber (Denk and Horstmann 2004; Hughes et al. 2014) or a focused ion beam (FIB; Chap. 6; Wei et al. 2012) is used to mill away the specimen. FESEM approaches have a significant advantage over TEM for examining larger samples and producing extended 3D data sets of larger tissue areas (i.e., 100s of microns volumes or several millimeters section slices). These types of dimensions are particularly suitable for more complex fungal tissues and fungal–host interactions, able to resolve individual cells and even organelles. For example, the 3D reconstructions of entire high-pressure frozen yeast cells by FIB/FESEM down to ~3–5 nm are isotropically resolved (Wei et al. 2012; Fig. 8.1). These images have significantly lower physical distortions as a result of material removal as compared to classical serial thin sections, and thus the associated 3D EM reconstructions align closely with the 3D light microscopy (LM) volume image (Narayan et al. 2014). Indeed, 3D EM data also can be directly and readily correlated with 3D LM, and X-ray micro computed tomography (Handschuh et al. 2013) and/or super-resolution images, as an extension to traditional correlative light and electron microscopy (CLEM) approaches (Caplan et al. 2011; Smith 2012; Perkel 2014).

8.4 Correlative Microscopy

Given that the power of any given method is often amplified when combined with other techniques (Smith 2012), we will focus on several advances in correlative microscopy. Combining information from two or more microscopic modes offers interpretive value that is greater than the sum of each data set alone. While fungal specific examples in the literature are relatively sparse, any of the microscopic methods discussed in this volume have the potential to be correlated and applied to modern mycology. Correlative microscopy can be generally classified into two categories: those that correlate data sets collected separately on individual microscopes, and those collected from fully integrated microscopy platforms capable of simultaneous imaging. If you have ever screened a sample on one microscope, typically with a large field-of-view and noted regions of importance prior to examination with a higher resolution microscope, then you have participated in basic correlative imaging. Indeed, the use of LM to screen samples for TEM gave rise to the now well-established term CLEM. The various microscopic modes have inspired myriad combinations of correlative microscopy as reviewed by Caplan et al. (2011). Here we describe new combinations and those we think are particularly applicable to mycology.

Multi-instrument Correlative Approaches Correlating data on the exact same region of a specimen using more than one microscope typically with the aid of masks, grids, and maps, requires highly skilled operators. As such, commercial solutions have been developed that seek to automate the tedious relocation steps (Perkel 2014; Smith 2012). Scanning electron microscopes (SEM), in particular field emission SEM (FESEM), are especially well suited to automation and have the advantage of allowing much larger serial sections to be collected on a single coverslip, glass slide, or silicon nitride wafer (Micheva and Smith 2007; Hayworth 2014). The use of silicon nitride substrates also facilitates successive TEM and tip-enhanced Raman spectroscopy (Lausch et al. 2014) that could serve as a new correlative approach for fungi.

Imaging fungal tissue structures such as fruiting bodies, plant–host interactions, or even larger scale colonies in culture could readily benefit from large-area imaging. The more automated approaches strategically place three fiducials on the substrate or sample holder for calibration on an LM. Relative coordinates are saved in the image file for rapid and precise relocation on a second microscope within minutes (i.e., FESEM). Such an approach was applied to collect structural illumination microscopy (SIM) and direct stochastical optical reconstruction microscopy (dSTORM) data using ultrathin sections of high-pressure frozen, freeze-substituted

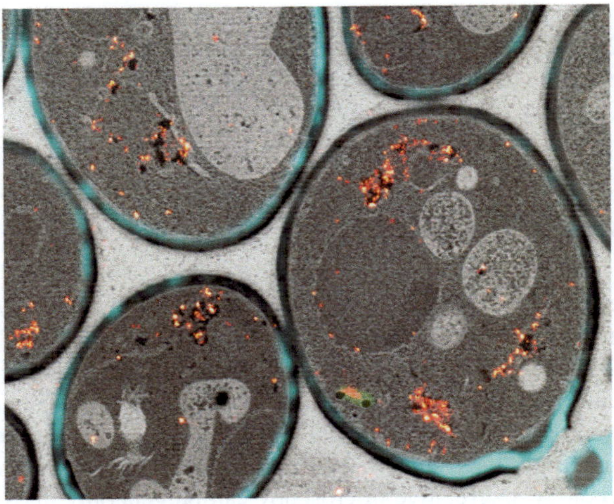

Fig. 8.2 CLEM using super-resolution with FESEM (Auriga Crossbeam, Zeiss) on thin sections of high-pressure frozen freeze-substituted *Saccharomyces cerevisiae*. Yeast cell walls *(blue)* labeled with Calcofluor white and Adenosin receptor *(red)* fused to Cerulean (hA1aR-Cerulean) and labeled with AlexaFluor®647 conjugated antibodies were imaged with structural illumination microscopy *(SIM)* and dSTORM localization microscopy, respectively, prior to additional heavy metal staining for FESEM. *BSE* images were acquired following automated recovery using Shuttle & Find software. Electron-dense cisternal compartments (Golgi equivalents) in FESEM image were directly correlated to hA1aR-Cerulean localization using dSTORM. (Image provided courtesy of Carissa Young, Jeff Caplan and Kirk Czymmek)

yeast samples embedded in HM20 resin (Fig. 8.2). Fluorophore labeled sections (AlexaFluor 647 anti-green fluorescent protein (GFP) antibody and Calcofluor counterstain) were then stained with heavy metals, moved to an FESEM, automatically relocated using correlative software and imaged using a BSE detector. Such an approach is fully amenable to widefield, confocal, and other optical microscopes. In particular, localization of single-molecule fluorescence on thin sections has some distinct advantages to traditional immunogold approaches, namely much higher labeling density and many more probes for targeting cell structure on a single section using Exchange-PAINT (Jungmann et al. 2014) or antibody stripping (Micheva and Smith 2007). Exchange-PAINT is a clever strategy that uses small, transiently binding fluorescent oligonucleotides coupled with affinity probes (biotin/antibodies) which can be multiplexed in series. Virtually an unlimited number of fluorophores can be used at each probe binding site for super-resolution single-molecule localization techniques (see Chap. 3).

Integrated Correlative Approaches Most modern microscopes have multiple optical components, for example, a typical LM will often be furnished with optics for bright-field, dark-field, phase contrast, or differential interference contrast (DIC). In this case, data correlation is straightforward since images are collected from multiple modes on a single region of the sample. Simultaneous multimode imaging requires more than one detector port and thus the full integration of two or more microscopes with pre-calibration or post-processing to ensure that the final images are in direct correspondence. Even for fully integrated microscopes with multiple detectors, the different time scales of each microscopic mode can present a challenge for simultaneous imaging and may also require time syncing of images.

With integrated LM/TEM approaches, thin resin sections are prepared for LM image acquisition within the TEM followed by sample reorientation and TEM imaging at the same location (Agronskaia et al. 2008). An elegant extension of this approach allows frozen/hydrated samples to be maintained at cryogenic temperatures within the TEM (Faas et al. 2013). Likewise, a modified optical microscope can be accommodated within an SEM chamber (Chap. 6) using high numerical aperture (oil immersion objectives), not possible with TEM approaches (den Hoedt et al. 2014). There are three major considerations when applying integrated CLEM approaches: (1) the impact of electron–sample interactions requires that organic fluorescent probe imaging precedes EM, since the electron beam can destroy fluorescence signal; (2) modification of labeling/staining strategies may be required to avoid quenching of fluorescence signals by proximal heavy metal stains; and (3) the impact of morphological distortions from changes in cell structure during sample processing. While integrated CLEM microscopes are not widely available, we do anticipate that this will change over the coming decade and prove to be particularly useful for correlative experiments on thin resin or cryo-sections of fungal specimens. Possibilities for cryopreservation, in particular, are intriguing to maintain high-fidelity sample structure and enable snapshots of living systems using integrated instruments such as LM/SEM (Kanemaru 2009), atomic force microscopy (AFM)/SEM (Williams et al. 2002), and AFM/microtome (Effimov et al. 2007).

The full integration of atomic force and optical microscopes is now mainstream for biological studies (Fig. 8.3). This combined correlative optical and physical scanned probe strategy permits targeting of desired cells, structures, and/or phenomenon with an optical microscope for subsequent high-resolution surface imaging, mechanical, and biochemical measurements (see Chap. 7; Kaminskyj and Dahms 2008). An early integrated scanning force-light microscope used surface imaging and optical sectioning with DIC to examine how the 3D dynamic cellular organization affects surface properties (Stemmer 1995). Combining these two instruments (Nagao and Dvorak 1998) came at the expense of each experiencing additional noise: vibrations impacted AFM imaging and red light from the AFM optical lever (Chap. 7) convoluted optical images. Optical microscope mounts that coupled vibrational noise into the AF microscope were eliminated with piezo-driven active vibration isolation units (Sandercock 1987), alongside replacing the red laser component of the AFM optical lever with an infrared (IR) laser (Gamble and West 1994). For biological applications, typically the AFM and its stage are seated on an inverted optical microscope which serves several purposes: aligning the detection laser on the AFM tip (Fig. 8.3a; Chap. 7; Fig. 7.1), previewing the sample (Fig. 8.3b, i.e., choosing a peripheral hypha from a mycelium or one yeast cell), and imaging gross phenotypic or morphometric features prior to high-resolution imaging (Fig. 8.3c).

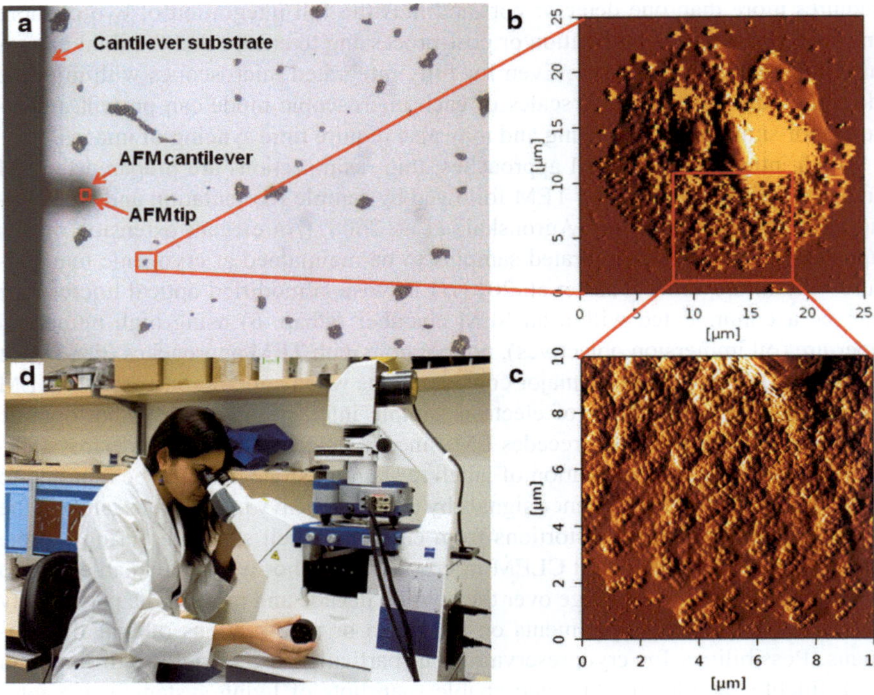

Fig. 8.3 Images of a *Candida albicans* deletion mutant exposed to 0.4 µg/ml caspofungin. **a** LM image (200×) of yeast on a glass substrate with *red arrows* indicating the out of focus AFM cantilever substrate, AFM cantilever, and AFM *tip* position. Corresponding AFM images at **b** *low* (200 × 200 pixels) and **c** *high* (500 × 500 pixels) resolution imaged in contact mode at 0.2 Hz with silicon nitride cantilevers (AppNano Hydra2R-50NG-10; nominal k ~ 0.084 N/m and radius of curvature < 10 nm) with *Z*-scale rendered in *brown (low)* to *white (high)* and fields of view as *red squares*. **d** Doctoral candidate Supriya V. Bhat examining her specimen (Bhat et al. 2015a, b) on a bright-field inverted microscope with *DIC* optics and modified condenser (Zeiss Axio Observer A1) housing an AF microscope outfitted with custom stage (JPK Nanowizard) all mounted on a piezo-driven vibration isolation platform (Herzan TS-150) and a small optical table (Halcyonics i4). Housed in the same room, on which the AFM can also be mounted, is a fluorescence microscope (Zeiss Axio observer Z1; λ_{ex} 365–580 nm) and a confocal laser scanning microscope (Ziess 780 with 458, 514, 488, 543, 594, 633 nm and pulsed two-photon Ti:Sapphire lasers) for multi-wavelength 3D imaging and FCS. (AFM images are courtesy of Spencer Zwarych and photograph by Geremy Lague)

Most commercial AFMs and optical microscopes can be integrated, with vendor choice mostly related to instrument features that satisfy user needs. A typical, relatively low-cost set-up (Fig. 8.3d) consists of an AFM outfitted with a custom stage mounted on a bright-field inverted microscope with fluorescence and/or DIC optics and modified condenser housed on a piezo-driven vibration isolation platform seated on a small optical table. The Dahms lab instrument in such a configuration (Bhat et al. 2015a, b) is capable of angstrom-scale resolution even without a sound isolation hood, despite its location on the fifth floor of their research building. Of course, the number of additional optical components is only limited by space and budget, and AFM instruments with portable electronic control units (ECU) can be mounted on multiple microscopic platforms. For instance, the Dahms lab AF microscope (Fig. 8.3d) remains in the configuration described above for routine imaging, but can be mounted on other nearby multiuser microscopes, namely an automated fluorescence microscope and a laser scanning confocal microscope (LSCM). The LSCM is equipped with both continuous wave and pulsed two-photon (Ti:Sapphire) lasers for multiwavelength excitation during confocal imaging (3D reconstruction) and fluorescence correlation spectroscopy (FCS; reviewed in Bacia and Schwille 2003), respectively. Any fluorescence or confocal (Chaps. 1–3) microscopy method can be simultaneously recorded with any AFM parameter (Chap. 7). The Dahms group is currently developing high-content atomic force confocal assays to probe changes in microbial phenotype, morphometrics (cell size/shape), cell wall ultrastructure and integrity (AFM; Ma et al. 2005, 2006; Kaminskyj and Dahms 2008; Jun et al. 2011; Paul et al. 2011; Bhat et al. 2015a, b) concurrently with intracellular physiological stress responses tracked by confocal (Chap. 1). Sample preparation methods developed for imaging growing hyphae (Ma et al. 2005; Paul et al. 2012) and yeast (Chap. 7) by AFM enable these high-content assays of live fungi.

High-resolution Fourier transform infrared spectroscopy (FTIR) mapping has found great utility in mycology (Chap. 4), but the spatial resolution and sensitivity have been limited by the relatively long wavelengths of IR light. Synchrotron infrared nanospectroscopy (SINS) overcomes this limitation by coupling synchrotron IR radiation with an AFM (Bechtel et al. 2014), enabling AFM imaging and near-field broadband IR spectroscopy with a tip-limited spatial resolution of < 30 nm (Berweger et al. 2013). This scattering scanning near-field optical microscopy (s-SNOM) technique focuses synchrotron IR light onto the oscillating tip of a modified AFM, located within an asymmetric Michelson interferometer. Detection of the elastically scattered light at harmonics of the tip oscillation frequency discriminates the near-field signal from the far-field background (Hillenbrand 2001). Movement of the reference arm mirror enables the weakly scattered signal to be frequency resolved and amplified by the stronger reference beam (Fig. 8.4a). Fourier transformation of the resultant interferogram produces a near-field spectrum (Bechtel et al. 2014) with a probing depth < 100 nm from the outer surface of the sample. The synchrotron source provides full mid-IR coverage from 700 to 5000 cm^{-1}, limited only by the choice of beamsplitter and detector. A high-resolution AFM image not only yields surface ultrastructural data but also yields a map to precisely localize the region of interest for FTIR (Fig. 8.4b). The SINS spectrum has signal-to-noise (S/N) ~ 100-fold better than standard FTIR microscopy. However, since this approach probes only ~ 30 nm into the cell wall in a 1 μm^2 area, as opposed to the entire cell, the realized S/N improvement is far greater (Fig. 8.4c). SINS of fungal hyphae at the Advanced Light Source (ALS) at UC Berkeley is currently limited by sample preparation (air drying) and data collection times for AFM (raster scan) and FTIR spectra, which could be overcome using an inline cryo stream, high-speed AFM (HS-AFM, Sect. 5) and more sensitive IR detectors, respectively. Despite its current limitations, the SINS instrument provides spatially resolved surface ultrastructure and chemistry of fungal cell walls (Fig. 8.4).

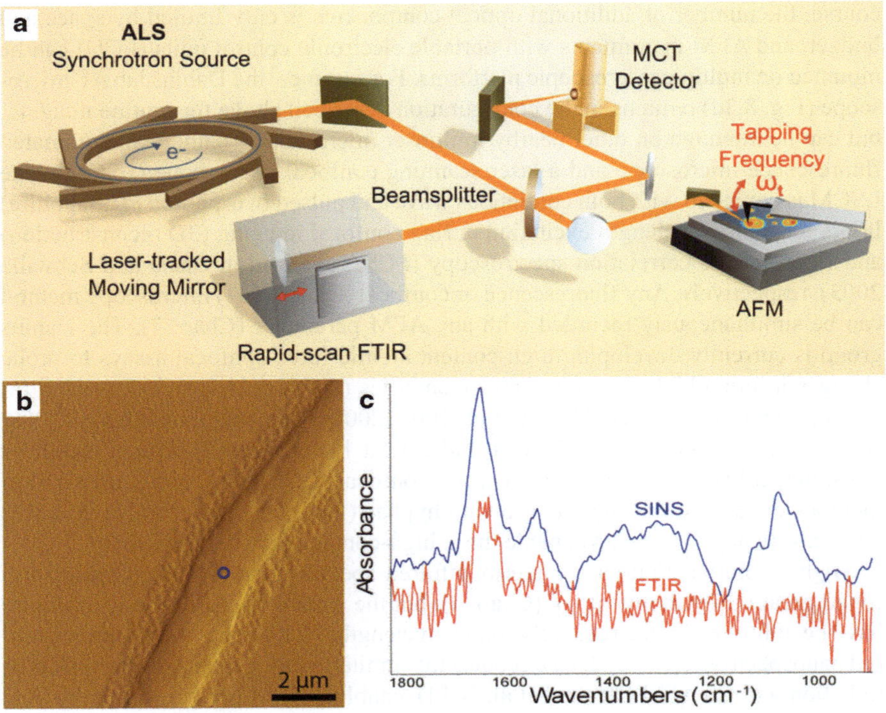

Fig. 8.4 **a** Schematic of the synchrotron infrared nano-spectroscopy *(SINS)* beam line at the 5.4.1 IR end station, at the Advanced Light Source, LBNL, Berkeley, CA, reprinted with permission from Bechtel et al. (2014). **b** AFM image of A1149 hypha approximately 15 µm behind the growing tip, acquired at the *SINS* end station. **c** *SINS* spectrum *(blue)* of the A1149 hyphal wall compared to a spectrum *(red)* of the same hypha using transmission with standard FTIR spectromicroscopy. Spectra were off set for easier visual comparison. The *SINS* spectrum shows a well-defined profile and approximately 100-fold improvement in S/N. The *SINS* spectrum in the near-field only probes the outer cell wall to approximately 30 nm depth, as evidenced from the high sugar content in the bands between 1200 and 1000 cm^{-1}, while the standard FTIR microscopy spectrum represents data from the entire cell, yielding a voxel of about 1 μm^3. The actual improvement in S/N is thus many orders of magnitude greater. (Schematic provided by Dr. Hans Bechtel, AFM image and FTIR spectra courtesy of Dr. Kathleen Gough)

Integration of super-resolution (Chap. 3) and other optical microscopes with AFM (Chap. 7) leads to intriguing possibilities. For example, combining AFM and stimulated emission depletion (STED) offers high-resolution intracellular fluorescence, topographical, and nano-mechanical imaging of cells (Chacko et al. 2013b; Harke et al. 2012) and cellular level nanomanipulation (Chacko et al. 2013a, 2014; Chap. 6). Stochastic optical reconstruction microscopy (STORM)-AFM has also been applied to study cell cytoskeleton (Chacko et al. 2013b) and surfaces (Zhao et al. 2014), and *in vitro* nucleic acid studies (Monserrate et al. 2014). Alternatively, combining AFM with synchronized confocal fluorescence lifetime microscopy (FLIM, Chap. 2), capable of tip-induced fluorescence quenching, yields fluorescence lifetime images that can be correlated with nanometer resolution (Schulz et al. 2013), which may allow for a more detailed understanding of hyphal development, and the impact of genetic deletion of antifungal targets or exposure to antifungal agents (Chap. 7). Using multiphoton excitation (Chap. 1) for correlative second harmonic generation along with photoacoustic modes for the label-free visualization of muscle and melanocytes in zebrafish larvae *ex vivo* (Shu et al. 2014) could prove very useful for examining fungal mycelia. When AFM is part of a total internal reflection fluorescence (TIRF) system, it can probe surface receptors (Lal and Arnsdorf 2010) with the potential for mapping fungal surface molecules.

Many hybrid microscopes are exclusively optically based. The adaptation of Raman microspectroscopy (RM) to fluorescence microscopy was used for correlative chemical analysis and fluorescence imaging of proteins and lipids in blood cells (van Manen and Otto 2007). This approach would have intriguing implications if applied to the work of Chang et al. (2010) exploring label-free spatiotemporal characterization of oxidative destruction of intraphagosomal yeast *in vivo* (autofluorescence imaging and RM). Similarly, label-free imaging approaches by Li et al. (2014) exploited RM with dark-field imaging in addition to fluorescence.

A number of optical imaging techniques have also been effectively combined with microfluidics to probe biological systems (reviewed in Wu et al. 2012), for example, exploring biofilms using microfluidic total internal reflection (TIR) and TIRF microscopy (Le et al. 2009). Examining biofilms by LSCM (Chap. 1) with confocal RM shows that lipids are key in biofilm formation across diverse phyla of bacteria from the environment, can define attachment phenotypes associated with various substrata, and identify key macromolecules involved in the attachment process (Andrews et al. 2010). The heterogeneity of matrix components and the function of multiple species in complex natural biofilms within their physicochemical microenvironment (Reuben et al. 2014) can be tracked with integrated LSCM (Chap. 1) and synchrotron radiation-FTIR microspectroscopy (Chap. 4). We believe these powerful correlative tools for understanding biofilm structure and biology could be readily extended to fungal biofilm investigations.

8.5 New Imaging Technologies

The synergistic development of biosensors and imaging technologies has resulted in the very high potential for new insights in mycology. Fluorescence imaging approaches, including confocal microscopy and epifluorescence imaging in their various forms (Chaps. 1–2), will continue to be fundamental and readily available tools for fungal biology into the foreseeable future. However, super-resolution microscopy (Chap. 3; Young et al. 2012) is ideally positioned to enhance our imaging and understanding of fungal molecules and structures. While there are numerous strategies to break the diffraction limit (Chaps. 2 and 3), we would like to highlight and emphasize two that could have exceptional promise for fungal studies, 3D localization (Baddeley et al. 2011) and Airyscan (Weisshart 2014) microscopy. Furthermore, we bring attention to two additional technologies, Bessel beam planar illumination (Planchon et al. 2011) and HS-AFM (Ando et al. 2013) which both greatly increase acquisition speed for photon- and scanning probe-based imaging, respectively.

Improvements in resolution are often accompanied by trade-offs that can limit the scope and kinetics of the biological question (i.e., requiring multiple image acquisition, significant processing time, and/or high laser powers). 3D localization microscopy is the highest resolution approach (~20 × 20 × 50 nm, *x-y-z*; Baddeley et al. 2011) with sub 10 nm resolution possible using improved probe strategies (Jungman et al. 2014). A common form of 3D localization microscopy requires that an asymmetrical point spread function (PSF) is produced by inserting an astigmatic lens or phase ramp into the light path (Baddeley et al. 2011). As a result, when single fluorescent molecules are imaged, the in-focus and defocused PSFs will have a characteristic shape relative to their position and distance from the focal plane. Since the PSF shape for each single molecule in the image is highly *z*-position sensitive, it can be modeled and a high-resolution 3D spatial map created (Baddeley et al. 2011). Typically, the rapid decline in signal, when imaging a single molecule in a diffraction-limited spot, leads to approximately 1 μm depth-of-field with 50 nm *z*-resolution for any given image. However, by taking 3D localization images at different *z*-planes, entire cells can be captured. Fixation is critical for super-resolution microscopy of fungi, as artifacts arising from movement (i.e., redistribution of cellular organelles/proteins/structures) are well known to be a particular challenge (Howard and Aist 1979). Fortunately, genetically encoded FPs for localization microscopy (Young et al. 2012) are compatible with freeze-substitution, minimizing these concerns. Indeed plunge-frozen freeze-substituted hyphal tips of *F. graminearum* expressing a calmodulin-mEos3 chimera have been imaged using 3D localization microscopy with multiple *z*-planes (Fig. 8.5). When FP fusions or genetic transformations are not possible, sample preparation using antibodies with confocal microscopy in freeze-substituted fungi (Bourett et al., 1998; Riquelme et al. 2002) could be slightly modified. Furthermore, the use of camelidae nanobodies, small (~3 nm) antigenic units derived from camels (Chavarty et al. 2014; Ries et al. 2012) would expand the applicability of 3D LM for fungi.

The addition of an Airyscan detector to an LSCM offers significant enhancement in resolution (1.7-fold over widefield) and sensitivity (Weisshart 2014). Instead of collecting a fluorescence emission signal from an LSCM at a single pinhole (see Chap. 1), it is focused on an array of 32 detectors. In confocal microscopy, decreasing the pinhole diameter improves resolution (and confocality), but at the significant cost of detection efficiency and S/N. With an Airyscan detector, while each detector element in the array is equivalent to 0.2 Airy unit (A.U.),[1] the entire array is effectively equivalent to collecting signal from a much larger 1.2 A.U. pinhole aperture

[1] Airy disk refers to the diffraction pattern of concentric circles surrounding a bright central spot created from a focused lens (Abbé 1873). The diameter of the first-order central Airy disc, objective lens dependent, is 1 A.U. The LSCM pinhole is typically set to an optimal signal-to-noise and resolution at ~1 A.U. Smaller pinholes (i.e., 0.2 A.U.) theoretically provide better resolution at the expense of signal-to-noise in a traditional LSCM setup (Chap. 1).

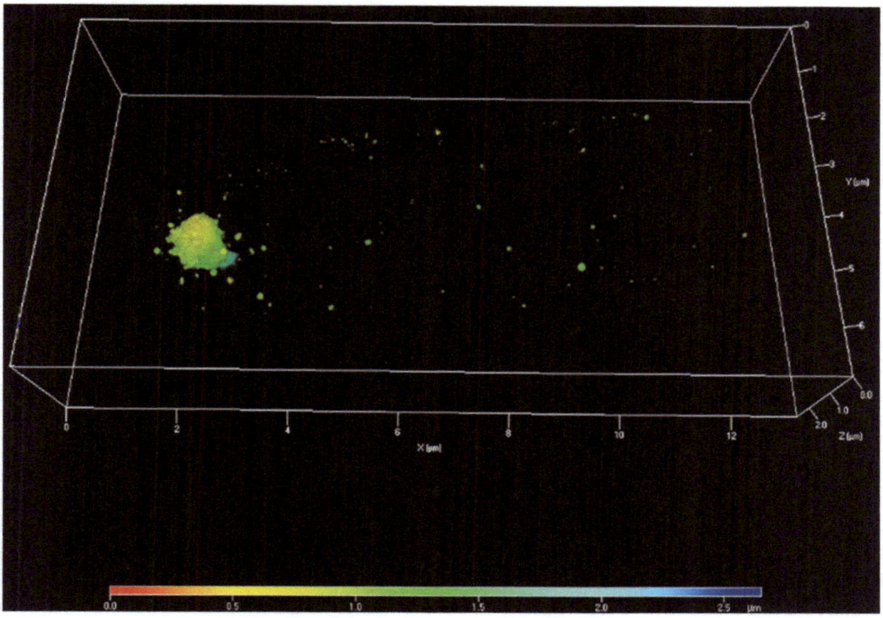

Fig. 8.5 3D photoactivated localization microscopy *(PALM)* shows super-resolution distribution of calmodulin in *Fusarium graminearum*. Expression of CaM-mEOS2 in *F. graminearum*, followed by plunge-freezing and freeze-substitution in aldehyde, revealed calmodulin localization at the Spitzenkörper core in the hyphal tip as well as numerous smaller puncta. Three-dimensional data acquired and rendered on a ZEISS ELYRA PS.1 at a xyz resolution of $\sim 20 \times 20 \times 50$ nm^3. (Image provided courtesy of Hye-Seon Kim and Seogchan Kang)

(see Fig. 8.6 in Weisshart 2014). In essence, additional resolution is gained by the individual smaller detector elements, effectively acting as very small pinholes. This resolution can be achieved without the signal loss of traditional LSCMs because signal from all elements can be utilized and reassigned from the entire 32-detector array with appropriate algorithms (see Fig B4 in Weisshart 2014). The relatively recent commercial introduction of Airyscan technology precludes reference to the current literature. However, preliminary Airyscan imaging of living *Saccharomyces cerevisiae* reveals clear improvement in S/N and resolution in mCherry and GFP-labeled membranes when compared to a single frame average LSCM (Fig. 8.6). Airyscan has clear advantages including increased scan speeds, little and no averaging and the ability to detect low-level fluorescent signals (i.e., endogenous expression). Unlike super-resolution techniques such as SIM and localization microscopy (Chap. 3), which require imaging in close proximity to the coverslip (i.e., 20 um or less for optimal results), Airyscan data can be collected deep within tissues (agar for fungal cultures). Indeed, this method excels under these light limiting, scattering conditions since the technology is less sensitive to the associated optical aberrations. As such, fixed or living samples traditionally intractable or very challenging for confocal or super-resolution based on weak signals, fast kinetics, or tissue depth (i.e., host–pathogen interactions) can be directly and readily imaged.

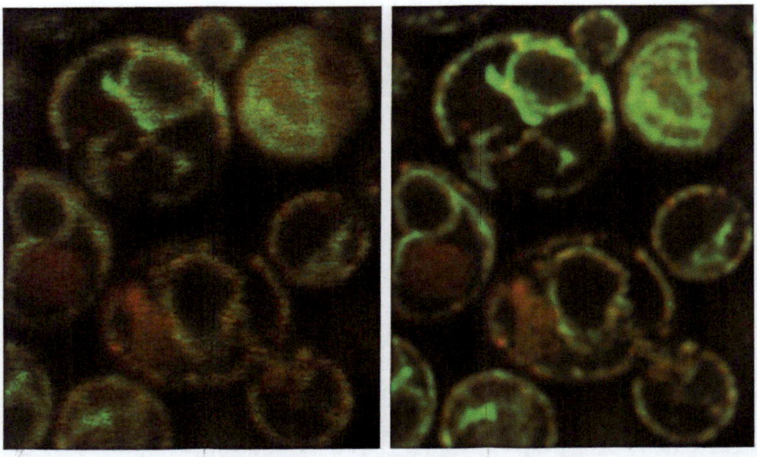

Fig. 8.6 Comparison of LSCM *(left)* and Airyscan imaging *(right)* of living *Saccharomyces cerevisiae* yeast cells. Live yeast cells expressing enhanced green fluorescent protein (eGFP) *(green)* and mCherry *(red)* fusion proteins both targeting the endoplasmic reticulum showed significantly improved resolution and S/N using the Airyscan array detector *(right image)* when compared to a traditional LSCM GaAsP detector *(left image)* when imaged under the same conditions. Data were acquired at the Zeiss Microscopy Labs, without averaging, on a ZEISS LSM880 with Airyscan detector. (Images provided courtesy of Carissa Young and Jeff Caplan)

As previously discussed, localization microscopy is often challenged by long acquisition and processing times that limit its utility for highly dynamic events. When kinetics is critical, other new tools deserve a closer look to help address those types of problems. Bessel beam plane illumination microscopy (Planchon et al. 2011) is one of a family of optical techniques that form a laser excitation sheet within the sample (Reynaud et al. 2015), thus creating an inherent optical section that illuminates an entire optical plane simultaneously rather than scanning a line or array of spots. While not fundamentally a super-resolution technique, this approach offers rapid imaging and improved sensitivity which lends itself to capturing full 3D volumes several orders of magnitude faster than most other optical sectioning methods. Bessel beam plane illumination has specific advantages over other plane illumination techniques as it is able to create much thinner planes and capture nearly 200 planes per second in the *z*-axis in living cells (Planchon et al. 2011). It can be expected that plane-illumination microscopy has the potential to fill a void in high-resolution and high-sensitivity 3D imaging of subcellular dynamics in living samples, including localization-based super-resolution microscopy (Zanacchi et al. 2011). Specifically, the potential to collect multiple, multicolor stacks every second from individual hyphal tips is very attractive for understanding the Spitzenkörper, polarisome, or anastomosis with unprecedented clarity.

Similar to how Bessel beam illumination accelerates 3D acquisition for LM, HS-AFM overcomes the relatively slow raster rate of scanning probe methods, typically requiring 30 s–30 min for high-resolution images (Chap. 7). The development of HS-AFM means that it is now possible to track changes in live cell mechanics and ultrastructure during dynamic processes. High speed with minimal impact was achieved by minimizing time delays in the feedback loop, dampening mechanical vibrations associated with fast scanning and developing more elaborate feedback control (reviewed in Ando et al. 2013). While performance relates to imaging conditions, HS-AFM can image fragile biological molecules in < 100 ms without structural or functional perturbation, allowing detailed mechanistic visualization in real time. High speed force mapping has been used to monitor cytoskeletal reorganization and growth, cell blebbing, and endocytic pit formation (Braunsmann 2014) and HS-AFM with exceptionally long and nm-sharp AFM tips has resolved filopodia morphogenesis, membrane ruffles, pit formation, and endocytosis in cultured cells at nm-s spatiotemporal resolution (Shibata et al. 2015). Although HS-AFM has yet to be applied to fungal cells, it has resolved the nanostructure of two Gram negative bacteria (Oestreicher 2015) and kinetics of fungal enzymes (Igarashi et al. 2012; Paslay et al. 2013; Shibafuji et al. 2014), demonstrating its applicability. For integrated correlative microscopes, HS-AFM enables simultaneous imaging with speeds more comparable to optical methods.

8.6 Conclusion

This arsenal of cutting-edge technologies is already generating new insights into complex biological systems, ushering in a new era of microscopic exploration. Concomitantly, the latest array of biomarkers, including sensors, FP fusions targeted to specific molecules of interest, and dyes targeting organelles, cell structures, and proteins, can offer unlimited combinations for adaptation to mycological systems. Unprecedented volumetric data from 3D microscopy, including electron, X-ray, and super-resolution, can be used to understand cell structure and function, developmental biology, and tissue architecture for fungal biology. Correlative microscopy, with sophisticated automated software for hybrid and multi-instrument approaches, offers multidimensional and multiscale information from different modes that can span an enormous field of view range (Å—mm). Ultimately, such approaches enable a holistic perspective of fungi at ultrahigh resolutions with novel insights about corresponding chemical, biophysical, and spatial relationships. Finally, the latest developments in optical and scanning probe microscopes vastly improve imaging time, enabling exploration of fungal phenomena in real time. We anticipate that these advances, if harnessed, will have immeasurable impact on our understanding of mycology.

Acknowledgments TESD thanks Spencer Zwarych for collecting the LM and AFM data, and Geremy Lague for photographing Supriya Bhat and the AFM setup shown in Fig. 8.3. Special thanks to Drs. Kathleen Gough, Catherine Liao, Hans Bechtel and Michael Martin and Susan Kaminskyj for the collaborative data collection, analysis and sample preparation for SINS at ALS (Berkeley), and to Dr. Gough for kindly preparing Fig. 8.4.

A very special thanks to Dr. Seogchan Kang and Dr. Hye-Seon Kim for permission to use the 3D super-resolution data, and Dr. Jeff Caplan and Dr. Carissa Young for their efforts and permission to use Airyscan and 3D FESEM/FIB images of yeast.

References

Abbé E (1873) Über einen neuen Beleuchtungsapparat am Mikroskop. Arch mikrosk Anat 9:469–480

Agronskaia AV, Valentijn JA, van Driel LF, Schneijdenberg CT, Humbel BM, van Bergen en Henegouwen PM, Verkleij AJ, Koster AJ, Gerritsen HC (2008) Integrated fluorescence and transmission electron microscopy. J Struct Biol 164:183–189

Akerboom J, Chen T-W, Wardill TJ, Tian L, Marvin JS, Mutlu S, Caldero NC, Esposti F, Borghuis BG, Richard Sun XR, Gordus A, Orger MB, Portugues R, Engert F, Macklin JJ, Filosa A, Aggarwal A, Kerr RA, Takagi R, Kracun S, Shigetomi E, Khakh BS, Baier H, Lagnado L, Wang SS-H, Bargmann CI, Kimmel BE, Jayaraman V, Svoboda K, Kim DS, Schreiter ER, Looger LL (2012) Optimization of a GCaMP calcium indicator for neural activity imaging. J Neurosci 32:13819–13840

Ando T, Uchihashi T, Kodera N (2013) High-speed AFM and applications to biomolecular systems. Annu Rev Biophys 42:393–414

Andrews JS, Rolfe SA, Huang WE, Scholes JD, Banwart SA (2010) Biofilm formation in environmental bacteria is influenced by different macromolecules depending on genus and species. Environ Microbiol 12:2496–2507

Bacia K, Schwille P (2003) A dynamic view of cellular processes by in vivo fluorescence auto- and cross-correlation spectroscopy. Methods 29:74–85

Baddeley D, Cannell MB, Soeller C (2011) Three-dimensional sub-100 nm super-resolution imaging of biological samples using a phase ramp in the objective pupil. Nano Res 4:589–598

Bechtel HA, Muller EA, Olmon RL, Martin MC, Raschke MB (2014) Ultrabroadband infrared nanospectroscopic imaging. Proc Natl Acad Sci U S A 111:7191–7196

Belousov VV, Fradkov AF, Lukyanov KA, Staroverov DB, Shakhbazov KS, Terskikh AV, Lukyanov S (2006) Genetically encoded fluorescent indicator for intracellular hydrogen peroxide. Nat Methods 3:281–286

Berweger S, Nguyen DM, Muller EA, Bechtel HA, Perkins TT, Raschke MB (2013) Nano-chemical infrared imaging of membrane proteins in lipid bilayers. J Am Chem Soc 135:18292–18295

Bhat B, Vantomme A, Yost D et al (2015a) Oxidative stress and metabolic perturbations in *Escherichia coli* exposed to sublethal levels of 2,4-dichlorophenoxyacetic acid. Chemosphere 135:453–461

Bhat B, McGrath D et al (2015b) *Rhizobium leguminosarum* bv. *viciae* 3841 adapts to 2,4-dichlorophenoxyacetic acid with "auxin-like" morphological changes, cell envelope remodeling and upregulation of central metabolic pathways. PLoS ONE 10:e0123813

Bourett TM, Czymmek KJ, Howard RJ (1998) An improved method for affinity probe localization in whole cells of filamentous fungi. Fungal Genet Biol 24:3–13

Braunsmann C, Seifert J, Rheinlaender J, Schäffer TE (2014) High-speed force mapping on living cells with a small cantilever atomic force microscope. Rev Sci Instrum 85:073703

Caplan J, Niethammer M, Taylor II RM II, Czymmek KJ (2011) The power of correlative microscopy: multi-modal, multi-scale, multi-dimensional. Curr Opin Struct Biol 21:686–693

Chacko JV, Canale C, Harke B, Diaspro A (2013a) Sub-diffraction nano manipulation using STED AFM. PLoS ONE 8:e66608

Chacko JV, Zanacchi FC, Diaspro A (2013b) Probing cytoskeletal structures by coupling optical superresolution and AFM techniques for a correlative approach. Cytoskeleton 70:729–740

Chacko JV, Harke B, Canale C, Diaspro A (2014) Cellular level nanomanipulation using atomic force microscope aided with superresolution imaging. J Biomed Opt 19:105003

Chakravarty R, Goel S, Cai W (2014) Nanobody: the "Magic Bullet" for molecular imaging? Theranostics 4:386–398

Chang WT, Yang YC, Lu HH, Li IL, Liau I (2010) Spatiotemporal characterization of phagocytic NADPH oxidase and oxidative destruction of intraphagosomal organisms in vivo using autofluorescence imaging and Raman microspectroscopy. J Am Chem Soc 132:1744–1745

Czymmek KJ, Bourett TM, Dezwaan TM, Sweigard JA, Howard RJ (2002) Utility of cytoplasmic fluorescent proteins for live-cell imaging of *Magnaporthe grisea* in planta. Mycologia 94:280–289

Czymmek KJ, Bourett TM, Howard RJ (2004) Fluorescent protein probes in fungi. In: Savidge T, Charalabos P (Imaging Meds) Methods in microbiology (microbial imaging), vol 34. Academic Press, New York, pp 27–32

Czymmek KJ, Bourett TM, Shao Y, Dezwaan TM, Sweigard JA, Howard RJ (2005) Live-cell imaging of tubulin in the filamentous fungus *Magnaporthe grisea* treated with anti-microtubule and anti-microfilament agents. Protoplasma 225:23–32

Denk W, Horstmann H (2004) Serial block-face scanning electron microscopy to reconstruct three-dimensional tissue nanostructure. PLoS Biol 2:e329

den Hoedt SV, Effting APJ, Haring MT (2014) The SECOM platform: an integrated CLEM solution. Microsc Microanal 20(Supp 3):1006–1007

Efimov AE, Tonevitsky AG, Dittrich M, Matsko NB (2007) Atomic force microscope (AFM) combined with the ultramicrotome: a novel device for the serial section tomography and AFM/TEM complementary structural analysis of biological and polymer samples. J Microsc 226:207–217

Faas FG, Bárcena M, Agronskaia AV, Gerritsen HC, Moscicka KB, Diebolder CA, van Driel LF, Limpens RW, Bos E, Ravelli RB, Koning RI, Koster AJ (2013) Localization of fluorescently labeled structures in frozen-hydrated samples using integrated light electron microscopy. J Struct Biol 181:283–290

Frommer WB, Davidson MW, Campbell RE (2009) Genetically encoded biosensors based on engineered fluorescent proteins. Chem Soc Rev 38:2833–2841

Gamble RC, West PG (1994) Scanning force microscope with integrated optics and cantilever mount. US Patent # 5291775 A, March 8

Handschuh S, Baeumler N, Schwaha T, Ruthensteiner B (2013) A correlative approach for combining microCT, light and transmission electron microscopy in a single 3D scenario. Front Zool 10:44. doi:10.1186/1742-9994-10-44

Harke B, Chacko JV, Haschke H, Canale C, Diaspro A (2012) A novel nanoscopic tool by combining AFM with STED microscopy. Opt Nanoscopy 1:3

Harris SD, Read ND, Roberson RW, Shaw B, Seiler S, Plamann M, Momany M (2005) Spitzenkörper meets polarisome: microscopy, genetics, and genomics converge. Eukaryot Cell 4:225–229

Hayworth KJ, Morgan JL, Schalek R, Berger DR, Hildebrand DGC, Lichtman JW (2014) Imaging ATUM ultrathin section libraries with WaferMapper: a multi-scale approach to EM reconstruction of neural circuits. Front Neural Circuits 8:1–18

Helmstaedter M, Briggman KL, Denk W (2008) 3D structural imaging of the brain with photons and electrons. Curr Opin Neurobiol 18:633–407

Hickey PC, Swift SR, Roca MG, Read ND (2004) Live-cell imaging of filamentous fungi using vital fluorescent dyes and confocal microscopy. In: Savidge T, Charalabos P (eds) Methods in microbiology (microbial imaging), vol 34. Academic Press, New York, pp 63–87

Hillenbrand R, Knoll B, Keilmann F (2001) Pure optical contrast in scattering-type scanning near-field microscopy. J Microsc 202:77–83

Hohmann-Marriott MF, Uchida M, van de Meene AML, Garret M, Hjelm BE, Kokoori S, Roberson RW (2006) Application of electron tomography to fungal ultrastructure studies. New Phytol 172:208–220

Honda A, Adams SR, Sawyer CL, Lev-Ram V, Tsien RY Dostmann WRG (2001) Spatiotemporal dynamics of guanosine 3′,5′-cyclic monophosphate revealed by genetically encoded, fluorescent indicator. Proc Natl Acad Sci U S A 98:2437–2442

Howard RJ, Aist JR (1979) Hyphal tip cell ultrastructure of the fungus Fusarium:improved preservation by freeze-substitution. J Ultrastruct Res. 66:224–234

Hughes L, Hawes C, Monteith S, Vaughan S (2014) Serial block face scanning electron microscopy—the future of cell ultrastructure imaging. Protoplasma 251:395–401

Igarashi K, Uchihashi T, Koivula A, Wada M, Kimura S, Penttilä M, Ando T, Samejima M (2012) Visualization of cellobiohydrolase I from *Trichoderma reesei* moving on crystalline cellulose using high-speed atomic force microscopy. Methods Enzymol 510:169–182

Imamura H, Nhat KP, Togawa H, Saito K, Iino R, Kato-Yamada Y, Nagai T, Noji H (2009) Visualization of ATP levels inside single living cells with fluorescence resonance energy transfer-based genetically encoded indicators. Proc Natl Acad Sci U S A 106:15651–15656

Jun AD, Signo K, Yost CM, Dahms TES (2011) Atomic force microscopy of a ctpA mutant in *Rhizobium leguminosarum* reveals surface property defects linking ctpA function to biofilm formation. Microbiol 157:3049–3058

Jungmann R, Avendaño MS, Woehrstein JB, Dai M, Shih WM, Yin P (2014) Multiplexed 3D cellular super-resolution imaging with DNA-PAINT and Exchange-PAINT. Nat Methods 11:313–318

Kaminskyj SGW, Dahms TES (2008) High spatial resolution surface imaging and analysis of fungal cells using SEM and AFM. Micron 39:349–361

Kanemaru T, Hirata K, Takasu S, Isobe S, Mizuki K, Mataka S, Nakamura K (2009) A fluorescence scanning electron microscope Ultramicroscopy 109:344–349

Kankanala P, Czymmek K, Valent B (2007) Roles for rice membrane dynamics and plasmodesmata during biotrophic invasion by the blast fungus. Plant Cell 19:706–724

Kim H-S, Czymmek KJ, Patel A, Modla S, Nohe A, Duncan R, Gilroy S, Kang S (2012) Expression of the Cameleon calcium biosensor in fungi reveals distinct Ca^{2+} signatures associated with polarized growth, development, and pathogenesis. Fungal Genet Biol 49:589–601

Lal R, Arnsdorf MF (2010) Multidimensional atomic force microscopy for drug discovery: a versatile tool for defining targets, designing therapeutics and monitoring their efficacy. Life Sci 86:545–562

Lausch V, Hermann P, Laue M, Bannert N (2014) Silicon nitride grids are compatible with correlative negative staining electron microscopy and tip-enhanced Raman spectroscopy for use in the detection of micro-organisms. J Appl Microbiol 116:1521–1530

Le NC, Yokokawa R, Dao DV, Nguyen TD, Wells JC, Sugiyama S (2009) Versatile microfluidic total internal reflection (TIR)-based devices: application to microbeads velocity measurement and single molecule detection with upright and inverted microscope. Lab Chip 9:244–250

Li H, Wang H, Huang D, Liang L, Gu Y, Liang C, Xu S, Xu W (2014) Note: Raman microspectroscopy integrated with fluorescence and dark field imaging. Rev Sci Instrum 85:056109

Löschberger A, Franke C, Krohne G, van de Linde S, Sauer M (2014) Correlative super-resolution fluorescence and electron microscopy of the nuclear pore complex with molecular resolution. J Cell Sci 127:4351–4355

Luo KQ, Yu VC, Pu Y, Chang DC (2001) Application of the fluorescence resonance energy transfer method for studying the dynamics of caspase-3 activation during UV-induced apoptosis in living HeLa cells. Biochem Biophys Res Commun 283:1054–1060

Ma H, Snook L, Kaminskyj S, Dahms TES (2005) Surface ultrastructure and elasticity in growing tips and mature regions of *Aspergillus* hyphae describe wall maturation. Microbiol 151:3679–3688

Ma H, Snook L, Tian C, Kaminskyj S, Dahms TES (2006) Fungal surface remodeling visualized by atomic force microscopy. Mycol Res 110:879–886

Martell JD, Deerinck TJ, Sancak Y, Poulos TL, Mootha VK, Sosinsky GE, Ellisman MH, Ting AY (2012) Engineered ascorbate peroxidase as a genetically encoded reporter for electron microscopy. Nat Biotechnol 30:1143–1148

Micheva KD, Smith SJ (2007) Array tomography: a new tool for imaging the molecular architecture and ultrastructure of neural circuits. Neuron 55:25–36

Miesenböck G, De Angelis DA, Rothman JE (1998) Visualizing secretion and synaptic transmission with pH-sensitive green fluorescent proteins. Nature 394:192–195

Miyawaki A, Llopis J, Heim R, McCaffery JM, Adams JA, Ikura M, Tsien RY (1997) Fluorescent indicators for Ca^{2+} based on green fluorescent proteins and calmodulin. Nature 388:882–887

Modla S, Caplan JL, Czymmek KJ, Lee JY (2015) Localization of fluorescently tagged protein to plasmodesmata by correlative light and electron microscopy. Methods Mol Biol 1217:121–133

Monserrate A, Casado S, Flors C (2014) Correlative atomic force microscopy and localization-based super-resolution microscopy: revealing labelling and image reconstruction artefacts. Chem Phys Chem 15:647–650

Nagai T, Yamada S, Tominaga T, Ichikawa M, Miyawaki A (2004) Expanded dynamic range of fluorescent indicators for Ca_{2+} by circularly permuted yellow fluorescent proteins. Proc Natl Acad Sci, U S A 101:10554–10559

Nagao E, Dvorak JA (1998) An integrated approach to the study of living cells by atomic force microscopy. J Microsc 191:8–19

Narayan K, Danielson CM, Lagarec K, Lowekamp BC, Coffman P, Laquerre A, Phaneuf MW, Hope TJ, Subramaniam S (2014) Multi-resolution correlative focused ion beam scanning electron microscopy: applications to cell biology. J Struct Biol 185:278–284

Nelson: http://onlinelibrary.wiley.com/doi/10.1111/j.1365-2958.2004.04066.x/abstract: Nelson G, Kozlova-Zwinderman O, Collis AJ, Knight MR, Fincham JRS, Stanger CP, Renwick A, Hessing JGM, Punt PJ, Van Den Hondel CAMJJ, Read, ND (2004) Calcium measurement in living filamentous fungi expressing codon-optimized aequorin. Mol Microbiol 52:1437–1450

Newman RH, Fosbrink MD, Zhang J (2011) Genetically encodable fluorescent biosensors for tracking signaling dynamics in living cells. Chem Rev 111:3614–3666

Oestreicher Z, Taoka A, Fukumori Y (2015) A comparison of the surface nanostructure from two different types of gram-negative cells: *Escherichia coli* and *Rhodobacter sphaeroides*. Micron 72C:8–14

Okumoto S, Jones A, Frommer WB (2012) Quantitative imaging with fluorescent biosensors. Annu Rev Plant Biol 63:663–706

Paslay LC, Falgout L, Savin DA, Heinhorst S, Cannon GC, Morgan SE (2013) Kinetics and control of self-assembly of ABH1 hydrophobin from the edible white button mushroom. Biomacromolecules 14:2283–2293

Paul BC, El-Ganiny AM, Abbas M, Kaminskyj SGW, Dahms TES (2011) Quantifying the importance of galacto-furanose in *Aspergillus nidulans* hyphal wall surface organization by atomic force microscopy. Eukaryot Cell 10:646–653

Paul BC, Snook L, Ma H, Dahms TES (2013) High-resolution imaging and force spectroscopy of fungal hyphal cells by atomic force microscopy. In: Gupta VK, Tuohy MG, Ayyachamy M, Turner KM, O'Donovan A (eds) Laboratory protocols in fungal biology. Springer, USA

Perkel JM (2014) Correlating Light and Electron Microscopy. BioTechniques 57:172–177

Pertz O, Hodgson L, Klemke RL, Hahn KM (2006) Spatiotemporal dynamics of RhoA activity in migrating cells. Nature 440:1069–1072

Planchon TA, Gao L, Milkie DE, Davidson MW, Galbraith JA, Galbrait CG, Betzig E (2011) Rapid three-dimensional isotropic imaging of living cells using Bessel beam plane illumination. Nat Methods 8:417–423

Reuben S, Banas K, Banas A, Swarup S (2014) Combination of synchrotron radiation-based Fourier transforms infrared microspectroscopy and confocal laser scanning microscopy to understand spatial heterogeneity in aquatic multispecies biofilms. Water Res 64:123–133

Reynaud EG, Peychl J, Huisken J, Tomancak P (2015) Guide to light-sheet microscopy for adventurous biologists. Nat Methods 12:30–34

Ries J, Kaplan C, Platonova E, Eghlidi H, Ewers H (2012) A simple, versatile method for GFP-based super-resolution microscopy via nanobodies. Nat Methods 9:582–4

Riquelme M, Roberson RW, McDaniel DP Bartnicki-Garcıa S (2002) The effects of ropy-1 mutation on cytoplasmic organization and intracellular motility in mature hyphae of *Neurospora crassa*. Fungal Genet Biol 37:171–179

Sandercock JR (1987) A dynamic antivibration system. Proc SPIE 0732, 1st Intl Conf on Vibration Control in Optics and Metrology, 157. doi:10.1117/12.937916

Sato M, Hida N, Ozawa T, Umezawa Y (2000) Fluorescent indicators for cyclic GMP based on cyclic GMP-dependent protein kinase Ialpha and green fluorescent proteins. Anal Chem 72:5918–5924

Schleifenbaum A, Stier G, Gasch A, Sattler M, Schultz C (2004) Genetically encoded FRET probe for PKC activity based on pleckstrin. J Am Chem Soc 126:11786–11787

Schulz O, Zhao Z, Ward A, Koenig M, Koberling F, Liu Y, Enderlein J, Yan H, Ros R (2013) Tip induced fluorescence quenching for nanometer optical and topographical resolution. Opt Nanoscopy 2:1

Shibafuji Y, Nakamura A, Uchihashi T, Sugimoto N, Fukuda S, Watanabe H, Samejima M, Ando T, Noji H, Koivula A, Igarashi K, Iino R (2014) Single-molecule imaging analysis of elementary reaction steps of *Trichoderma reesei* cellobiohydrolase I (Cel7A) hydrolyzing crystalline cellulose Iα and IIII. J Biol Chem 289:14056–14065

Shibata M, Uchihashi T, Ando T, Yasuda R (2015) Long-tip high-speed atomic force microscopy for nanometer-scale imaging in live cells. Sci Rep 5:8724

Shu X, Lev-Ram V, Deerinck TK, Qi Y, Ramko EB, Davidson MW, Jin Y, Ellisman MH, Tserevelakis GJ, Soliman D, Omar M, Ntziachristos V (2014) Hybrid multiphoton and optoacoustic microscope. Opt Lett 39:1819–1822

Smith C (2012) Two microscopes are better than one. Nature 492:293–297

Stemmer A (1995) A hybrid scanning force and light microscope for surface imaging and three-dimensional optical sectioning in differential interference contrast. J Microsc 178:28–36

Subach FV, Subach OM, Gundorov IS, Morozova KS, Piatkevich KD, Cuervo AM, Verkhusha VV (2009) Monomeric fluorescent timers that change color from blue to red report on cellular trafficking. Nat Chem Biol 5:118–126

Terskikh A, Fradkov A, Ermakova G, Zaraisky A, Tan P, Kajava AV, Zhao X, Lukyanov S, Matz M, Kim S, Weissman I, Siebert P (2000) "Fluorescent Timer": protein that changes color with time. Science 290:1585–1588

Tsien RY (2011) A genetically encoded tag for correlated light and electron microscopy of intact cells, tissues, and organisms. PLoS Biol 9:e1001041

Tsutsui H, Karasawa S, Okamura Y, Miyawaki A (2008) Improving membrane voltage measurements using FRET with new fluorescent proteins. Nat Methods 7:683–685

van Manen HJ, Otto C (2007) Hybrid confocal Raman fluorescence microscopy on single cells using semiconductor quantum dots. Nano Lett 7:1631–1636

Varnai P, Balla T (2007) Visualization and manipulation of phosphoinsitide dynamics in live cells using engineered protein dynamics. Eur J Physiol 455:69–82

Wei D, Jacobs S, Modla S, Zhang S, Young CL, Cirino R, Caplan J, Czymmek K (2012) High-resolution three-dimensional reconstruction of a whole yeast cell using focused-ion beam scanning electron microscopy. Biotechniques 53(1):41–48

Weisshart K (2014) The basic principle of airyscanning, pp 1–11. ZEISS

Williams PA, Papadakis SJ, Flavo MR, Patel AM, Sinclair M, Seeger A, Hesler A, Taylor RM, Washburn S, Superfine R (2002) Controlled placement of an individual carbon nanotube onto a microelectromechanical structure. Appl Phys Lett 80:2574–2576

Wu J, Zheng G, Lee LM (2012) Optical imaging techniques in microfluidics and their applications. Lab Chip 12:3566–3575

Xu X, Gerard ALV, Huang BCB, Anderson DC, Payan DG, Luo Y (1998) Detection of programmed cell death using fluorescence energy transfer. Nucleic Acids Res 26:2034–2035

Young CL, Raden DL, Caplan J, Czymmek K, Robinson AS (2012) Optimized cassettes for live-cell imaging of proteins and high-resolution in yeast. Yeast 29:119–136

Zaccolo M, De Giorgi F, Cho CY, Feng L, Knapp T, Negulescu PA, Taylor SS, Tsien RY Pozzan T (2000) A genetically encoded, fluorescent indicator for cyclic AMP in living cells. Nat Cell Biol 2:25–29

Zanacchi FC, Lavagnino Z, Donnorso MP, Bue AD, Furia L, Faretta M, Diaspro A (2011) Live-cell 3D super-resolution imaging in thick biological samples. Nat Methods 8:1047–1049

Zhang C, Czymmek KJ, Shapiro AD (2003) Nitric oxide does not trigger early programmed cell death events but may contribute to cell-to-cell signaling governing progression of the *Arabidopsis* hypersensitive response. Mol Plant Microbe Interact 16:962–972

Zhao W, Tian Y, Cai M, Wang F, Wu J, Gao J, Liu S, Jiang J, Jiang S, Wang H (2014) Studying the nucleated mammalian cell membrane by single molecule approaches. PLoS ONE 9:e91595

Index

T. E. S. Dahms and K. J. Czymmek (eds.), *Advanced Microscopy in Mycology*, Fungal Biology, DOI 10.1007/978-3-319-22437-4

MIX
Papier aus verantwortungsvollen Quellen
Paper from responsible sources
FSC® C105338

If you have any concerns about our products,
you can contact us on
ProductSafety@springernature.com

In case Publisher is established outside the EU,
the EU authorized representative is:
Springer Nature Customer Service Center GmbH
Europaplatz 3, 69115 Heidelberg, Germany

Printed by Libri Plureos GmbH
in Hamburg, Germany